Feline Cultures

ANIMAL VOICES
ANIMAL WORLDS

Robert W. Mitchell, series editor

Feline Cultures

CATS CREATE THEIR HISTORY

Éric Baratay

Translated by Drew S. Burk

The University of Georgia Press

ATHENS

Published with the generous support of the Institut Universitaire de France.

Athens, Georgia 30602
www.ugapress.org

Designed by Kaelin Chappell Broaddus
Set in 11/13.5 Corundum Text Book Roman
by Kaelin Chappell Broaddus
Printed and bound by Sheridan Books
The paper in this book meets the guidelines for permanence and durability
of the Committee on Production Guidelines for Book Longevity
of the Council on Library Resources.

Most University of Georgia Press titles are
available from popular e-book vendors.

Printed in the United States of America
24 25 26 27 28 P 5 4 3 2 1

Library of Congress Cataloging-in-Publication Data

Names: Baratay, Éric, author.
Title: Feline cultures : cats create their history / Éric Baratay ; translated by Drew S. Burk.
Other titles: Cultures félines (XVIIIe-XXIe siècle). English
Description: Athens : The University of Georgia Press, [2024] | Series: Animal voices, animal worlds | Includes bibliographical references and index.
Identifiers: LCCN 2023051930 | ISBN 9780820365145 (paperback) | ISBN 9780820366593 (epub) | ISBN 9780820366609 (pdf)
Subjects: LCSH: Cats—History. | Domestic animals—History.
Classification: LCC SF442.6 .B3713 2024 | DDC 636.8—dc23/eng/20240207
LC record available at https://lccn.loc.gov/2023051930

Originally published in French under the title
Cultures félines (XVIIIe–XXIe siècle): Les chats créent leur histoire

CONTENTS

Feline Cultures

The Cat Massacres of the Eighteenth Century

—. . . so, a q-u-i-c-k j-u-m-p . . . rEsTlesS ears RAISED THEN Lowered fur RAISED tail WhiPPinG THEN LOWERed HEAD Down . . . Intrigued, worried FRIGHTENED . . . GROWING CRIES aPPROACHING . . . known humans . . . odors, gestures, threateNING soUNDS . . . evasion . . . PAINFUL Cries . . . Smashed, o-b-s-tr-uc-ted . . . panic effort SUFFERING . . . Explosion . . . —

Jerome had delivered the final blow to the skull. Léveillé, who had put an end to the creature's flight "with one quick blow" to the kidneys, just as quickly gathered up La grise and tossed her "into the first gutter in sight." The two typography apprentices of a Parisian printer in the 1740s are living under poor conditions: hardly any interest from their artisanal master, a cook who has taken to selling part of their food rations under the table to someone else and serving them instead the rations left for the house cats, the additional chores (along with the accompanying reprimands) required of the workers in the early morning hours make the arrival of night drag on even longer. Such a scenario was fairly common for apprentices, but the critical way in which these two young men accept their task only leads to more resentment and violence: "The masters love the cats and so they should therefore take to hating them." Nevertheless, they do not expect to suffer the consequences of this "murder that one must keep concealed."[2]

It's true. La grise is adored by the lady of the house, but the text by Nicolas Contat, revealing the details of the act twenty years later through the character of Jerome, leads one to believe that La grise is not the only cat: there are also other cats whose presence is tolerated under the rooftop of the printing workshop or whose presence is appreciated for keeping the mice out. The cat's name—La grise—"grey," serves as a way to distinguish her from the others, to indicate she is the favorite without being unique and that her way of life is the same as the others, coming and going as she pleases, leading the lady of the house to often be out looking for her "everywhere." And yet, the cat's position within the community of the printing shop seems to have elicited a singular relation with the workers and the apprentices to whom she often pays a visit more than the other felines. They are figures whose paths she must cross often while they're in the midst of working long hours. And the cat has learned to distinguish and recognize each of them by way of their voices, odors, and movements, since a cat has such abilities, through which she is capable of establishing a cold proximity:[3] *none of these men seem to want to seek out contact. And it was such a disposition that led her to not be so suspicious on one particular day, that is, to not be able to truly grasp their intentions, improperly evaluating the noise, their postures, and their approach, realizing all too late and allowing herself to be easily captured and without the need for any sort of trap (as was the case with the other cats).*

For Jerome and Léveillé are also seeking their vengeance against perhaps several (specific?) cats whose nocturnal cries above their shelter had prevented them from sleeping and acquiring "a short break" from the "persecution" and "pains" of the daily grind.[4]

> It's true that the nocturnal cries of cats were the object of a myriad of complaints during this era and that such disturbances were well documented in France at least until the beginning of the twentieth century. In attics and gardens and on rooftops, receptive female cats would call out to male cats through "piercing, reverberating screams" whose intensity was capable of waking an entire "regimen of monks," to which one could also add the roars of male cats fighting and the long moaning of eventual mating partners whose calls one "disliked to hear the most when in bed" and that were all the more "discordant" and "disagreeable" given that humans were un-

able to hear the entire spectrum of calls.[5] Such disturbances were the result of a strong presence of city cats tolerated in Western cities in order to hunt rodents, left to their own devices in order to learn to fend for themselves, and who reproduced at will. Given the permeability and similarity of lifestyles and the territories in which they resided, such populations were reinforced by other felines considered more or less to be house cats.

The racket caused by the cats leads to a "jealousy for both Monsieur Jerome and Léveillé; they take it on themselves to not be the lone unfortunate souls and seek to welcome the master and mistress to accompany them in their despair." For three consecutive nights, Léveillé climbs up on the roof and imitates the cat calls, leading the proprietors to ask of the boys "if they couldn't figure out a way to get rid of these destructive animals," of course, while making sure to keep La grise. However tempered the request, it is taken by the two apprentices as carte blanche and a free-for-all to "hunt down" the cats and to rally "a part" of the "workers" to the idea of social vengeance by way of the cats, or as a form of "festive" entertainment. Sacks are strung up on corners of the gutters, hunting beaters are positioned at entries to stores, and Léveillé, having climbed up on the roof to hide the body of La grise—the first cat to be killed—"tosses the little terror onto the neighboring gutters."[6]

—. . . JoLT, "Ear-tURN," BLoODcuRDLing screAMS muffled SOUNDS . . . paws on the ground treMORs . . . "puPILS" aGiTaTions slipping SLIDING . . . afraid, FEAR FRIGHTENED . . . s-c-a-m-p-e-RUNFLEE . . . leave enter avoid leave again bypass circle around . . . sudden stumble instantaneous darkness compressed turned-over tangled-up body . . . paNIC, meOWs SUDden STarts CLaW ScraTCHES . . . sacks dredged bodies raised thrown . . . PAIN back paws SmASHEd head . . . neck GRabbed RaiSed . . . MeOWing mOVing . . . meN, gESTURES oDORS SCREAMS ALL ARoUnD . . . squeezed neck SCREEAM TWIST CLAAWW . . . SQUEEZED NECK STIRRED . . . SQUEEZED NECK . . . heavy body . . . —

For, after having knocked out a couple of felines and tossed them in a knapsack, the apprentices and workers both organize some "entertainment" for "some laughs": a parody of a trial, with judge, guards, confessor, and ex-

ecutioner in order to condemn and hang the rest of the cats. Alerted by the cries and agony of the remaining cats, the lady of the house accuses them of having killed La grise, which they mischievously deny while the master of the house yells at them for having fun instead of working. Both the lady and the master, however, can sense the social tension ("these wicked men can't kill the masters; they killed my cat") and quietly withdraw with prudence while leaving the cats to hang while the workers, "who love disorder," "are in a state of great joy" to such an extent that Léveillé will perform this scene of the bosses twenty times in order to "ridicule them," for which he will receive "resplendent applause" since "all of the workers are banded together against the Bosses, all it takes is to speak ill of them in order to be esteemed." Then the tension returns from the shadows and everyone gets back to work.[7]

> In a renowned work, the historian Robert Darnton has uncovered the human details of this *Great Cat Massacre*, most notably the tensions inside the printing workshop and various other Parisian workshops of the period: horseplay and other antics during communal parties, such as the carnival and its mayhem, or during professional gatherings such as Saint Martin of the printers, punctuated by parodies of various bosses, but also veiled critiques in particular concerning the severity of the print master or the supposed adultery on the part of the mistress juxtaposed along with symbolic scapegoats for whom she must have lusted in the same manner or that unduly captured one's attention such as, in this case, the felines.[8] We will only focus on the feline side of the story.

Besides La grise, the various additional cats' strong suspicions of others make it so that traps are required, no doubt due to the cold distance and indifference they are accustomed to maintaining in their daily encounters with men, because of which they are not accustomed to any punctual routines of disturbance as they lived alongside humans at a distance and not with them. But there were occasions when such an existence gave way to confrontation. *Moreover, the print master and his wife lament merely the presumed demise of La grise and not the survival of the other cats, whose death they don't dare investigate out of fear for their own self-preservation and the fact that cats can easily be replaced. They did what a lot of cat owners among the bourgeoisie and aristocracy of the period did (perhaps "owners" is too strong a word here; "hosts" would be the better term), who went to great lengths to separate their house cats from those living in the attics and on the rooftops who were considered much more common, despicable, and contemptable.*[9] *If the workers don't differentiate be-*

tween the cats and don't understand the passion bestowed on some of them, and if the master and lady of the house don't approve of the violence toward them, they all share nonetheless a certain depreciation of the cat that is well expressed by the actors or at the least emphasized by Contat. Jerome would have considered the cats screaming on his rooftops as "wild or bedeviled" and the print master and his wife would be persuaded, following the first night of Léveillé's imitations, that one must keep one's distance from such "wicked animals," perhaps even alert the parish priest.[10] *So, it's not by chance that the mock trial with a confessor staged by the workers makes one think of such trials that were also once performed against witchcraft.*

Such a reading is deeply rooted within Western Christianity. Whereas the Bible and the Fathers of the Church make no mention of the cat, a connection was invented between the cat and the devil during the twelfth and thirteenth centuries. Such connections were made during battles with the Vaudois and above all in fighting against the Cathars, a term that was just as quickly claimed to be a derivation of *catus* (the term for "cat" in Latin), which began appearing in the fifth century but whose true origin is unknown. The Cathars were heretics accused of worshipping Satan in the form of a cat, and in so doing, they were said to have renewed ties with ancient cults and their sexual perversions. Medieval and modern art have placed the cat in religious scenes in order to represent the Devil opposing God, evil versus good, the fall in contrast to salvation. At the same time, fairy tales, legends, and novels assured the cat was endowed with mysterious powers of predicting the weather, the arrival of visitors, illnesses, misfortune and death (indicated through various cat gestures), all of which were believed in by the masters of the printing workshop, convinced and frightened as they were of an impending curse.[11]

This connection with evil assures the diffusion of a poor portrait of the cat that is hardly praiseworthy (the cat is wild, mean, clever, a hypocrite, ungrateful), culminating in Buffon's writings from 1756, which bestowed even more credit to such stereotypes given his renown as a naturalist and his famous work *Histoire naturelle*. The cat is deceitful and perverse, with the "obliquity of their motions and the duplicity of their looks." He is nothing more than a tease for man, making him think that the cat has an attachment to him but which is merely nothing more than pure appearance. The cat shows himself to be unruly and disloyal. He observes, exploits, and plunders with malice. Furthermore, he is cruel and destructive in regard to other weaker animals.[12]

And now, dear readers, I must pause this tale in order to point out something of great importance. If you were to com-

> pare your cats to the portrait mentioned above, you would be engaging in an anachronism, and if shocked by such portrait, you would be merely indulging in its partiality and falsehood. There is no need to convince you that Buffon was merely collecting a list of fantasies and stereotypes. In reality, his writings present a human *reading* and *translation* of a feline situation that is still ongoing. Beyond the shortsightedness of the words and moralizing judgments of the naturalist, one can indeed glean the conditions and behaviors of cats common to the era: of cats in the public gardens, living in attics, on rooftops, and in residences, which we can also see noted in documents, such as those of Contat. In other words, Buffon does not provide us with a portrait of the *Cat as such* or of *our cats*, but rather of the cats of the eighteenth century. The entire philosophy found throughout the pages of *Feline Cultures* holds onto the premise of the historical variation of human discourse *and* that of feline behaviors.

These common cats have the right to move freely anywhere and are left to their own independent movements but are never provided aid, are hardly ever solicited or named, and are rarely or poorly fed (to prevent them from becoming lazy). They are aware of and know their environs (and the other animals therein) better than any of the neighboring humans, who remain cold, distant, and sometimes threatening; they evolve with prudence, suspicion, always prepared to slink away, to such an extent, Buffon writes, that "we hardly ever encounter them"; they are first and foremost concerned with their subsistence, always prepared to wander, prowl, and hunt into the wee morning hours. Always on the lookout, ready to seize on their prey. What Buffon has to say in his writings is quite similar to what we can imagine taking place in the print workshop: "Those most familiar are not under any subjection, but rather enjoy perfect freedom, as they only do just what they please [. . .]. Besides, most of them are half wild, know not their masters, frequent other granaries, and never visit the kitchens and offices belonging to the house but when pressed to it by hunger."[13] It's only through taking on the animal-situation that the interpretation-deformation-transformation is undertaken, that disapproval is hammered down by a human society that values the submission of its populace to a human being bearing a divine authority (the king among humankind, the human among other animals) and loyal groups as opposed to fickle individuals. Such an apparently paradoxical attitude (condemning the marginality of an already marginalized group) leads to the very reinforcement of suspicion, perpetuating the distance from these cats and further solidifying their utilitarian role. Buffon considers that having a feline merely in order to "amuse oneself" is a "misuse" and that their "use"—namely the norm—is for hunting.[14]

> I'd like to now warn you of this game involving the formation and consolidation of feline situations. The dark portrait painted of the cat imposes a condition onto these individuals and therefore onto their rather singular behaviors. Cats merely reinforce such a portrait while around humans and as a result reinforce the behaviors of the humans. The latter confront the behaviors of the felines, and so it goes . . . Human representation thus plays an important role in the cat environment. An environment that determines the cats' condition and behavior. Their behavior is not some fruit of an intemporal nature; rather, it's the product of a specific situation and environment. And their behavior will change (within the limits of biology) in relation to any change in their situation and environment. Please keep in mind that the objective of this book is to demonstrate that the various ways of being a domestic cat (*Felis catus*; I'll leave aside for the moment wild/feral cats, *Felis sylvestris*) greatly fluctuate under the influence of humans.

Largely shared throughout all social classes, such a pejorative judgment elicits uncontrollable fears that have since been forgotten and that today we would find surprising and on par with the contemporary phobias of snakes and spiders. La Bruyère recounts how "Béryll fainted at the sight of a cat," and Moncrif shares another example with us in an aristocratic salon where everyone criticized his defense of felines, when a "cat appeared, and the first thing that occurred was the immediate disappearance of one of my adversaries; everyone became angry with me." As for the presence of cats in literary histories, in England, Samuel Johnson's biographer Boswell confesses his antipathy toward felines and a certain malaise when the writer's own cat is present. Moreover, Johnson even goes to great lengths to attend to his own errands so as not to "disgust" his assistants with the presence of his cat.[15] Hence the commonly shared conviction that, through a mixture of fantasy and unfounded fears, a danger resides with cats. Moncrif tells us that "one has heard since the cradle that Cats are treacherous by nature; that they suffocate infants; that perhaps they are sorcerers." The latter suspicion will appear again in the writing of Contat, when he writes that "the bedeviled cat performs a sabbath all night long" on his rooftop, whereas it's actually Léveillé who imitates the cries of cats and who "will perform his sabbath and would be mistaken for a sorcerer if one did not know him." Already beginning in the thirteenth and fourteenth centuries, direct allusions had been made between witches, evil, and cats. Witches were said to always be surrounded by cats and participate in sexual saturnalia in sharp contrast to proper values of the day, making use of their fur as ingredients in their evil spells, even transforming into cats to effectuate their misdeeds; it was also said that cats were themselves sorcerers or demons. Such fantasies and beliefs will lead Moncrif to note, concerning one well-known scientist of the period, that "Monsieur de Fontainelle avows that he was raised to believe that on the eve of St. John's Eve there remained not a single Cat in the Cities, because on that day they betook themselves to a general sabbath." Such ideas only helped to reinforce the practice of burning cats on this very same day of June 23 as a kind of defensive gesture, which actually happened in cities such as Metz in the Lorraine region where the notion that cats could transform into witches was widely held, and where a great number of women were burned at the stake between the fourteenth and seventeenth centuries.[16]

— . . . sudDENLY. . . puPiLS quick forward movements restless . . . "QUIVER-SNIFF" STRONG ODORS . . . s-c-u-r-r-y flee . . . turn, swerve . . . FlAtten . . . neck pulled RAISED UP . . . FEAR ANGER ROAR "sCRaTCH" . . . tossed darkness "pupil-diverting-DILATE" cats below . . . meow, push claw cats above below alongside . . . TReMOr . . . MIxTures TReMoRS, cries, scratches, bites, rOUNDed backs FOLded BaCK Ears . . . scrEAMS human ODORS ENveloPING rising FIdgETinG ScRUM . . . STOP . . . heat crackling below DeSCENt . . . BURns FEAR JumP MEOW . . . DESCeND BURNS PaIN FEAR JUMP CRY . . . DESCeND SWELLING SuffERING SCREAMS FIRE . . . —[17]

Yelling, laughter, applause: pleasure. "A true pleasure can be seen by the people captivated by the screaming and meowing of the cats as they near the flames," notes the Benedictine Jean François at the beginning of the 1770s. For the inhabitants of Metz, the spells were most likely warded off, cast away, and banished thanks to these events. This would explain the presence of the military garrison defending the city and the county magistrate charged with preserving the populace from plagues, who would "gather en masse with a serious demeanor." The latter, having doubtless purchased cats from trackers, were the ones to set the logs ablaze thanks to the "fire makers" perched on top of the basket full of cats. Did military officers and municipal workers truly believe in such stories about witchcraft in the eighteenth century? The answer to that question remains up in the air, for the major-

ity of intellectuals and social elite had already dispensed with witch hunts beginning in the second half of the seventeenth century. And the same could be said for the inhabitants of Metz. François refuses to connect such an incident with witchcraft as it would no doubt lead to the city and its elites being enveloped in ridicule. He merely leaves such a notion to the crowd ("one must be the people in order to believe") and posits another more banal cause: "One laughs at such things: that is enough in order to perform such acts." Nevertheless, it was quite possible that a number of notable individuals actually believed cats to be harbingers of misfortune, of signs of possible ills to come, and would therefore like nothing more than to banish their presence. Others perhaps merely wanted to adhere to this old "custom," which still pleased and relieved many to the extent that most people didn't really care as to the fate of the cats. And even though in one of his manuscripts he describes them as unfortunate beasts and requests, in another publication, that they should be left alone, François only reserves his condemnation of such "juvenile pleasures" as being a part of "human stupidities," along with other "bizarre" ceremonies, such as the celebration of the pagan rite of summer solstice, and merely muses that it's a waste of firewood that would be better off reserved for the poor![18]

Such incidents in Paris and Metz are nothing more than examples of the widespread hatred and violence that permeated the Middle Ages and the Enlightenment. Cats are obviously not the only animals to have suffered such aggressions; nevertheless, their omnipresence, their sheer numbers—along with their affordability and the bad press surrounding them—predestines them to such dismal fates more than others. Poorly reported, such daily brutality of the era is difficult to clearly measure, which makes Contat's written description all the more valuable. For a long time this kind of brutality had been justified thanks to the negative symbolism we have already mentioned and the condemnation of any sort of affection, foregrounded in the various stories of hermits who felt God's wrath for having shown affection to felines, in dictates warning of the dangers of being in their proximity, in the sharing of faulty banter whereby it's always children, women, or the elderly who appreciate them and whose opinions are considered as unreasonable. Familiarity is made to rhyme with hardness: "There's always an occasion to beat one's cat," declares an ancient medieval proverb.

And so it is that in a good number of other locales, cats are burned at the stake for cause of witchcraft or for pleasure, or they're simply cast off belfries and belltowers, put to death as a game or as a result of mass hysteria, or to assure the prosperity of men and cultures during religious or civic festivals, or in a specific episodic manner peculiar to certain areas. Cats are even buried inside the initial construction and scaffolding of buildings as a preventive measure intended to keep rodents at a distance. Felines are also part of the fur trade. Skinned cats are used to produce fat or medicinal potions. Their capture and killing is all the more accentuated during times of famine, war, or sieges, when hunger leads to seeking out the most familiar impure animals such as cats, mice, or even more valuable animals such as dogs or horses. It's during hard times like these that we see the invention of discreet declarations found in the song for instance by Michel Mère in the eighteenth century, or in the old rumors about the use of cat or dog meat in restaurants.[19]

And yet, in 1773, Marie-Charlotte de Seneterre, a young Parisian aristocrat who, three years earlier, had married the marshal of Armentières—the governor of Metz—will request from the municipality "grace for the cats" and will subsequently receive a declaration definitively reprieving their burning at the stake. Metz was one of the last big cities in the West to still practice such punishments. In Paris, for example, such a practice had already been outlawed by the king at the beginning of the eighteenth century. This would eventually lead to a decline in the practice of cat burning, but it still remained common in rural areas during certain periods throughout the year until the nineteenth century. For the nineteenth century would usher in a return to an appreciation of cats, which had already begun during the 1600s.[20]

Already during the Middle Ages, a handful of documents exist indicating a certain sensibility toward the beauty of and contact with the common cat, as well as a more concrete affection through a well-honed attention to them, including the rare practice of even granting the cat a name. Often such careful attention to cats was practiced by women of the aristocracy, monks, and nuns who withdrew into the contemplative and intimate life of a chamber or monastic cell. Among aristocrats and writers, the interest in certain cats becomes all the more affirmed during the Western Renaissance. One of the most important factors being the import of Syrian and then Persian cats sought for their beauty and precious rarity, whose coats often were of a brilliant white—the antithesis of the vulgar black cat of the gutters. These princes, aristocrats, or bourgeois, who often bore a hatred for local cats, began to promote their cats as animals made to be looked at: stroked and caressed, beribboned, perfumed, furnished with necklaces and pillows, allowed to roam in living areas and salons, bred among themselves so as not to become debased. The breeding of cats, along with surveillance of their reproduction, care, and selection, really begins to take off in Europe during this time period in these milieus. This interest in cats is also accompanied by the first introduction of a literary appreciation of them, however minor it may be. In France, it can be seen in such writings as du Bellay's epitaph for his cat Belaud (1558) and the work *Chats*, written by Moncrif (1727). The latter, himself an aristocrat, published his book on cats in order to promote the role they had played throughout the history of humanity, such as their role in Egyptian cults, as well as their presence alongside individuals of power and intellect, from Montaigne to Richelieu. Literary portraits, depicting a graceful and intelligent cat, with a heightened sensibility, delicate and bestowing gratuitous affection, are sketched out by the author, who counts such aristocrats and writers as his principal readers.[21] During the same era, art begins to move from using the cat as a symbol to the cat as individual. To be sure, one can still find depictions of daily scenes that represent the animal as a symbol—to signify for example, passionate love—but many other depictions only evoke an intimate presence. Such a preference is reinforced through artistic depictions of children and women holding cats or playing with them, containing a sentimental dimension that began to be depicted in the 1750s. Starting in the 1760s, one even begins to see such sentimental depictions in portraits specifically focused around cats and their daily life. This inflexion of representations, which concerns Western Europe to varying degrees (it appears most pronounced in England), does not lead to the creation of cats as the house pets or companions that we recognize today.[22] The relation promised in eighteenth-century literature remains rather loose, more in line with Contat's depiction of the cat La grise mentioned above. Moncrif, for instance, doesn't even think it necessary to name his favorite cat. "One needs only to call him by his name: *Cat*, one simply says to him." He also declares that he doesn't believe it necessary to look after their well-being: "They lavish on us the grace of their society. When we receive them into the intimacy of our families, they do not try to play there any role but that of animals; they demand none of the attentions which men owe only to men, and spare us the shame of reckoning among our occupations the duty of satisfying their needs & caprices."[23]

> Carefully consider the aforementioned quote. Beyond a mere politeness and distinguished humor—claiming to grant to cats the same free choice as one would leave to any invited guests—allowing for their approval of any imposed proposition, it suggests that there is not only a human influence on felines but also an influence of the latter on the former,

with reciprocal effects. Here, cats raised as such will behave as such, and they determine or reinforce human dispositions as well.

During this period, the difference between common human behaviors and new behaviors resides above all in the refusal of violence toward animals. Nevertheless, the turn in representations leads to a partial but growing tendency, largely in the minority during the eighteenth century but ever increasing throughout the nineteenth and twentieth , which has since become the majority attitude (today, violence exerted toward cats is considered scandalous and is met with subsequent punishment). We see a progressive modification of human behaviors toward cats, including to their environment and as a result, to their machinations. Today, cats have become our companions, our pets, and are the most appreciated animal companions in Western countries as well as in a Westernized Japan. They now outnumber dogs, whom they dethroned at the beginning of the twenty-first century, and have become more and more interactive and cooperative with humans.[24]

Thus, I have chosen a chronological expanse between the eighteenth and twenty-first centuries, throughout which different, parallel or successive environments are deployed in the West, so as to present to you this variability of domestic cat behaviors within time and space, as well as the grounding tendency, not so much systematic but sensitive: the increasing pressure on humans toward more proximity and interactions with cats, leading these felines, according to Buffon, to a state demonstrating "less attachment to persons than to houses," and more recently to the inverse.[25]

I therefore want to argue against the notion of the domestic cat having an unchangeable nature. I also shall dispel a certain portrait of cats that has become commonplace: of cats being mysterious, independent, and unpredictable, which in the past led them to be hated and which since then has led us to love them. Far too many people forget that this depiction, sketched out by naturalists, writers, artists, and philosophers—most notably in the nineteenth century—is inspired by a very particular humano-feline situation, which secreted a very specific feline environment and condition leading to the production of very specific cats who embody a version of the cat, but not that of *the Cat* as such. Beyond this species, it becomes a question of disproving the belief in an unchangeable permanence of animal behaviors that would only develop by way of constant biological drivers: an instinct, a drive, a genetic capital . . .

I recognize that the behavioral plasticity of the domestic cat is now accepted by ethologists, ecologists, and veterinarians. But they often limit the scope of such plasticity, limiting it to tiny psychological variations, miniscule ecological adaptations here and there, fairly weak or ephemeral within a present conflated with the past or the future and therefore considered as intemporal. My colleagues have not (yet) grasped—but it's certainly true that it's up to historians to demonstrate it—that behavioral flexibility also takes place within time. That, for example, there is enough variation demonstrated between an eighteenth-century cat that would become anxious due to any close encounter with humans (who were often hostile) and the behavior of the contemporary catdogs who quickly become anxious any time their attentive owners take leave of them. Such variations should help us to confirm the existence of a history of cat behaviors. In other words, following my reasoning, you should no longer believe in the immutable essence of an *Eternal Cat*; rather, you should believe in historical cats! You will see the construction, diffusion, and transformation of feline attitudes and their fluctuation within time, through adaptation to changing environments. And such a fluctuation and adaption implies a dynamic and an arbitrary relativity of individual behaviors. It also implies a diversity of cats in space, since modifications oscillate in their intensity according to place, adaptations varying according to individuals and groups.

But perhaps you are already beginning to wonder: just what exactly is the reason for the different dispositions of your paragraphs? For I'm inviting you on a historical promenade with four different levels: that of cats (written with Times New Roman), that of humans (written in italics), the historical context (the body of the text written in a smaller font and right justified), and that of scientific reflections (same body of smaller font, but with center justified paragraphs), in order to distinguish each of them while also showing how they are intertwined.

In fact, let's return to the feline context—or rather the humano-feline context, since humans have become more and more influential in its development throughout the centuries thanks to the various frameworks and attitudes they impose on cats. The rise in an appreciation of cats has provoked a variety of situations, conditions, and ways of being of feline cultures—from the most ancient and constant such as stray cats, to the most recent such as catdogs, with an ever-increasing gap within time and space. However, up until the twentieth century, all the various feline situations and ways of being shared the common trait of being tied to the territories frequented. Contrary to what many believe in connecting the domestic cat with nature, this is not a natural state, since our current era has demonstrated the possibility of different behaviors. Rather, it's a historical situation due to the rejection, suspicions, and distance (however well-meaning) manifested by humans to such an extent that cats practice . . .

PART 1

A Withdrawal into Territorial Cultures

. . . that it is possible to uncover through the gathering of archival documents: a forest territory for the cats who wandered in the woods, a countryside territory for cats who wandered in the fields, a farming territory for cats whose territory was the farm, an urban territory for their wandering vagabond colleagues of the streets and the village or city rooftops, a neighborhood territory for the felines who were more or less attached to a specific neighborhood or city center, and a residential territory for those cats more or less confined to apartments. This territorial repartition partly corroborates that established by ethologists regarding the cat's relation to humans: independent cats—referred to as stray cats in this book (referred to by ethologists as *feral cats*, but which conflates them with their wild cousins); vagabond cats (which ethologists refer to as *semiferal*—the same remark about conflation could be made here) who oscillate between autonomy and a human refuge; farm cats who are fixed to a locale but who are responsible for feeding themselves; and felines belonging to a certain property who are more or less free to roam as they please around an abode or who are kept inside.[1]

The first three territorial situations—forest, countryside, and farm— have only recently been understood through studies bearing on our current era.[2] As far as the past is concerned, the available documents are rare or nonexistent: humans from these various places did not write and writers in the past were rarely interested enough to write about them. With the individual approach taken up in this present book, it is impossible to have an exacting classification regarding the probable modifications in relation to such cultures. I will therefore focus on the three other areas—the urban cityscape, the residential neighborhood, and the domicile—for which fairly consistent documentations do exist and which will serve to create the wide brushstrokes of quasi-photographic moments, or embodied portraits. Such dispositions are rare throughout the nineteenth century; they will see an increase throughout the twentieth century and have now become abundant over the past thirty years following the rise in popularity of cats, except for the small number of . . .

CHAPTER 1

Street Cats

. . . hardly evoked at all, until very recent inquiries.[1] However, there are a couple of exceptions, most notably the literary work of Paul Léautaud (1872–1956), who worked for the publishing house Éditions Mercure de France. Léautaud was a writer in his free time and the author of a voluminous journal only published beginning in the 1940s, wherein he notes a myriad of accounts concerning abandoned or stray dogs and cats for which he expresses a great pity in the same manner as for all those layers of excluded humanity. Such accounts by Léautaud can be found above all between 1908 and 1912, during a brief period when he decides to dedicate himself to animals while still residing in a Parisian apartment, where he can only welcome as many animal friends as his then-current significant other will tolerate. He therefore decides to take up caring for the stray animals directly in the streets on the narrow cobblestone paths of the Odéon neighborhood where he works. Entire pages of his journal describe him pacing and scouring the sidewalks in search of stray cats, seeking to provide them aid. Such a task becomes an obsession for Léautaud and takes up endless hours and days, an obsession that Léautaud doesn't want to truly present within the pages of his edited journal, which explains why a large portion of his writings on animals would only be posthumously published much later as *Bestiaires*.[2] Within these pages, he becomes the true reporter of the streets, describing feline situations, as well as the attitudes of other animals and humans. Only evoking what he captures with his own eyes, he makes no mention of the cats residing on the rooftops or in the attics (moreover, the decrease in complaints at the beginning of the twentieth century regarding the nocturnal cries of the rooftop cats leads one to believe that their actual numbers had already diminished through eradication) and only focuses his attention on those cats on the ground, in the courtyards, living in enclosed areas and other terrains enmeshed alongside the streets, within visible and accessible spaces. What Léautaud is able to report does not allow for individualized portraits (he is not present enough to follow one specific cat, nor do his remarks contain adequate descriptions), but merely a collection of photographed groups composed of beings . . .

Wandering, Independent

(Paris, 1908–1912)

. . . several of which (a "dozen," "several") live in a permanent location within the enclosed courtyard and gardens *of a bathhouse or shuttered seminary.*[1] Nothing is mentioned of their daily life, *which is of no interest to the people surrounding them.* It is quite possible that the indigenous cats and newcomers to such spaces have carefully taken their time to decipher and assimilate to them, slowly gaining a mastery of them. In order for this to happen, the cats make use of their hearing, which is much better than that of humans, carefully detecting any noise or low-frequency sounds such as the ultrasound made by vermin, and instantly pivot their ears to amplify the source of the noise. They are also capable of smelling an abundance of odors, be these distant, or from the past, whether changing or dissipating, coming or going, past or present. All of this allows the cats to live within four dimensions—overlaying scents within space, and odors within time—scents from the past, present, and future. In the end, cats see a much larger panorama than humans, but it is a flat panorama (containing only two dimensions—height and length). They are only capable of deciphering movements and being alerted a possible danger. Such a panoramic vision allows them to quickly turn their head to see what's ahead of them. Allowing them to now see in relief and granting them the capacity to locate the increasing or decreasing disturbances, motionless presences, and other obstacles and reliefs in order to choose a route while modifying and quickly adjusting to the variations of luminosity. Dilating or contracting their pupils in order to flee or scurry off to different spaces, be they shaded or out in broad daylight. Cats are just as adept at venturing out at night or at dusk, when their prey is on the move, as they can detect the weakest levels of luminosity. Cats also possess an acute way of distinguishing forms, sizes, and textures, enabling them to differentiate a variety of objects, living beings, and individuals, the slowest being the least easily located while those closest to them would remain within a fluid temporality but would betray their location because of their scent.[2]

> I borrow all these descriptions from the contemporary cat, as it's safe to assume their ancestors perceived in the same manner a century ago. However, we should keep in mind the

possible objection to such similarities in light of recent understandings provided by epigenetics, which we will return to later on: namely, the epigenetics around the modulation of one's capacities and intensities in relation to one's specific environment. We've already begun to understand the role of epigenetics, for example, in the case of soldiers who fought in World War I heralding from cities that were better illuminated at night, who, as a result, were unable to see as easily at night in the trenches as their colleagues who heralded from the countryside where night remained much darker. Therefore, it's plausible to think that such sense faculty on the part of cats was more developed in cats living on the streets, because they were subjected to environmental pressures. This means that we must think and construct a history and geography of perceptions and physiologies! A vast program that I will set aside for now, but which, it should be clear, opens up the possibility of variation in all its facets.

Certainly, the cats of past centuries marked their territories with urine and above all with personal substances such as facial pheromone F3, which is secreted thanks to a subcutaneous gland dispersed from head to tail while passing through paws and hindquarters. The cats then secrete scent by rubbing against objects or prominent vegetation. In so doing, they establish an olfactory map helping them to situate and orient themselves, to indicate their presence in a stronger or weaker register depending on their temperament, and to subsequently manage an area by transforming it into a territory.[3] No doubt, having lived for a long time in groups, they learned upon their arrival or at a very young age to share these gardens among themselves, modifying the territories in accordance with the evolution of the group, to separate them, juxtapose them, gather them together, or even merge them in the event the need arose for strong cooperation.

You will have perhaps noticed that I speak about territory and not about space. Behavioral ecology qualifies life space as one's ecological dwelling, varying according to one's physical environment, available food, climate, etc., because this ecology thinks at the scale of species or groups of specific types of cats. Within such an optic, ecology conceives the territory as an exclusive space that is protected in order to satisfy one's vital needs. As I'm focusing on cats as singular actors and beings, even when they are within groups, it's better to envision territory in another manner. In order to do so, I shall take

> inspiration from geographers, who insist on repartitioning, sharing, and zones of encounter between actors. Or anthropologists, who focus on individual or collective ways of how place is appropriated and then constituted as a territory.[4]
> Constructed for thinking about humans (for complex beings), these approaches can also be used to conduct research into cats, since we now know and recognize more and more the complexity and wealth of intelligence innate to cats. Hence our recourse to the social sciences [*sciences humaines*], which I will present to you little by little. But only by avoiding any engagement in an anthropomorphism, while considering the biological, cognitive, and behavioral specificity of cats, and thereby de-anthroposizing the concepts that, as we shall see, arise out of the social sciences.
> The basic thing to retain here is that such a reading will allow us to open up our reflection to a wider and more diverse group of readers. As such, the term territory opens itself up to the question of appropriation, that is, of individual, social, and cultural variables. Ecologists insist more and more on the variation of the vital domains of contemporary cats.[5] But they are willingly open to changing the factors behind such variation (food, population density, forms of human habitat, seasons, etc.) in terms of felines considered as hardly autonomous. Whereas we should be asking how cats are integrating such givens and act in accordance with them, but each in their own personal way, according to their initial acquired skills, according to their unique know-how in transforming a place into a territory.

Free to roam in the gardens deserted by humans, these cats more than likely communicate with each other. And here as well, they more than likely behave like their contemporary peers by using their sensoria to express themselves and decipher each other. They therefore must adopt accentuated, clear, and distinct postures in order to be understood and seen: raising up on their feet, arching their back and bending their tail in order to express their anger or to cuddle, their head withdrawn to the shoulders to express fear or their tail straight up so as to signal appeasement or friendliness. They doubtless rub their heads and hindquarters between themselves in order to indicate their social insertion, engage in a harmonious encounter, perhaps even exchanging substances in order to share their state of being (such as the facial pheromone F4, which translates a good intention and which would reduce any sort of aggression), perhaps even creating a friendly group scent. This would lead each cat to often sniff the others, first and foremost sniffing each other's heads, the most prominent bearer of one's individual and perhaps collective identity. These cats would rarely need to meow in order to express a demand, a need for another cat or partners, or rivals, and would purr among relatives or other cats with whom they were familiar in order to express quietude.

Nowadays, research into cat colonies tells us that cats are very sociable, especially in tiny areas such as this. Hardly any hierarchy is practiced, and even less any sort of aggression toward each other. It's quite the opposite, as they often demonstrate a profound intimacy: the related mothers, having grown up together, collaborate; other individual cats share their favorite places with other cats, without, however, hunting for food together. The whole varies in intensity depending on the size of the groups and more than likely the environments and individuals.[6] We can suppose that such instants of solidarity must have also taken place in *our aforementioned Parisian gardens of the early twentieth century*. Moreover, Léautaud perceives cats "playing and frolicking." Such scenes would perhaps not take place in another area *depicted by the writer*: an "*enclosure with tools on the Boulevard Raspail, in the middle of the road, right were the street intersects with the rue de Fleurus," temporarily under construction*, where "several" strays have decided to make their home, no doubt in order to seek a bit of peace and quiet behind the construction barriers; but they don't know each other, unless some of them arrived there together and had already been living together before in cooperation.

The relations documented in these two cases by Léautaud are significantly different. One being a *"closed off" area alongside the street where one would find a pregnant cat along with a male and two kittens no older than five or six weeks*, and, on the other hand, a courtyard for a residential building where, "it would seem, a tiny kitten awaits the arrival of his mother cat who only passes through occasionally to nurse him." Rather than being kittens adopted by other cats, something which would be rare for cats who were living in complete freedom, the cats at the construction site were more than likely the kittens of the mother but who must have endured an early separation due to the mother becoming pregnant again much too soon. Such a situation can be explained by the fact that some female cats can be fertile again only two weeks after giving birth, even while still lactating, and by the daily presence of the male cat. We now know that female cats are often solicited from two different angles and are therefore often undernourished, tired, even to exhaustion from trying to care for several offspring while also being solicited by other abandoned male kittens, and that this can lead them to become irritable, less motherly, and even lead to the early weaning of her litter. For it's the mother who takes care of the litter and not the male cats who, often, have already been quite skillful beggars and are more than likely meowing all around her.

In this case, in the courtyard of the residential building, based on the male kitten's attitude to the female cat attending to his needs, we're dealing with a kitten who has been neglected by his mother, a mother who had been isolated and was now hardly a mother and more preoccupied with survival; therefore, the male kitten was also undernourished and certainly meowing just as much as the mother. With problems of growth, a certain early death would soon follow for the kitten, which explains even today the small percentage of male cats that end up being strays. Male cats run the risk of learning difficulties when left on their own too soon, having lost the early opportunity to emulate their mother, mostly notably for hunting, which immediately reduces their chances for overall survival as well as being detrimental to their social development. Such lack of hunting ability leads to the male kittens to become fearful and aggressive, which subsequently limits both their integration into a feline group and their ability to become close with humans.[7]

One of the cats in the fenced off construction area bites Léautaud, and the other cat disappears. The cat in the courtyard is "extremely wild and impossible to approach." In fact, the cats in the courtyards and gardens behave similarly to those on the street ("inapproachable, they immediately flee and are impossible to find, *as soon as we open the gate* to the courtyard") in spite of their situational differences and doubtless differences in character: they are certainly more socialized with their fellow cats but only within these deserted areas, less so with humans. They all share the same reaction:—suddenly . . . quiver-sniff . . . distant odors . . . increasingly intense . . . different from self of those nearby . . . unknown . . . "pointed-ear" ear-turn . . . loud noise . . . movements along the side of "turned-head" mass in front . . . afraid fear terror scurry flee ears tail lowered fur . . . hidden flattened . . . on the lookout . . .—

Everywhere, the adult cats act in a similar manner, thereby reinforcing and conserving this attitude: the cat couple from the construction site scurries away; one male, "so frightened in his abode made of wooden planks" escapes as soon as Léautaud tries to "*approach him by talking to him*," not knowing, like the others, how to decipher *human intentions*, unable to differentiate themselves from these other perceived beings and judging them to be identical. Cornered, the cats become aggressive: the female cat at the construction site, having sought refuge on an adjacent street in a basement windowsill, struggles away when *the writer attempts to gather her up*: the male—leaps, SeTs HiS pAwS In MoTIoN . . . FALLING INTO a hole [into another basement windowsill in a nearby park] . . . BRUshING up AgAinST . . . carefully scrutinizing CAsTs Out his paw [against the hand passing between the bars on the window] sCrATch SCrATch, obliging Léautaud to give up. We can now discern that such reactions are not instinctive or innate but inherited, since it would seem that certain repetitive responses to stress are transmitted, with the aim of a preventative adaptation, from the mothers to the child in the fetus, which will later on become fearful adults.[8] At the same time, such a heritage is also an acquired behavior, created or reinforced by every generation through the observation and imitation of the mothers who remain alongside their kittens, as well as through a reinforced activity within the environment, in this case, the human environment.

In seeking to perturb the "homeless" cat couple with two kittens who sought refuge in the enclosed construction site alongside the street, a "maid from the neighboring building takes pleasure in tossing them pieces of coal in hopes of eradicating them." From the cobbled area of the site, workers nearby can be seen casting bits of wood onto the cat's tiny shelter.

> A number of other felines experience similar instances of daily violence. Léautaud encounters a number of cats overwhelmed by projectiles sent from concierges attempting to rid the area of their presence or sent hurling by local boys having a bit of fun. Such is their fate when the cats aren't simply killed, like the poor cat smashed in the street, or the cat that was hanged and whose eyes were poked out, with his body pierced open by numerous knife wounds, leading to a slow agonizing death. The writer assures the members of the SPA (Société protectrice des animaux) that other cats are simply tossed alive into furnaces at the École des Beaux Arts by med students, or attacked by dogs and strangled close to the wine merchants of Les Halles, or skinned alive for the eventual sale of their hides or for their meat, sold under the table. With the overriding sentiment by the directors of the Père Lachaise cemetery that there are too many cats

living on the grounds, a number of cats are snared within makeshift traps and simply tossed, dead or alive, into the cemetery's crematorium. Others, in the Luxembourg Gardens, find themselves crammed into cages and drowned in a city pond. Through a discreet arrangement with the institutions that purchase them on the side, cats also find themselves being swallowed whole by snakes at the Jardin des plantes or dissected at the university.[9]

Mere exaggerations of a misanthrope? Between the nineteenth and first half of the twentieth century, literature abounds with evocations and confessions (from Woolf to Joyce in English letters, and from Flaubert to Genevoix by way of Zola, Proust, Gide, and Camus as concerns French letters) detailing beaten, mutilated, and burned cats, most notably harmed by children from the streets and countryside seeking to amuse themselves.[10] Such violence is a result of a persistent depreciation of cats in spite of a return toward the admiration by the elites. Even among the latter, documents suggest that such violence is still prevalent at least until the 1860s. One example comes from the writer Tounessel who, in 1847, calls for "ridding the country of such thieves" with the following example: "I never encounter a cat out on the hunt in the countryside or the woods without firing a rifle shot in his direction." Even though this sort of devaluation of cats' lives will diminish in the second half of the nineteenth century, the violence and prejudice of the recent past still has a strong influence on the common values of the day in how cats living on the streets are treated—as the most underappreciated of all.[11]

Threats and violence stoke, perpetuate, and reinforce the cats' fears. The cat living in the Jardin des bains passes the sidewalk and is spotted *by four people who want to catch and kick him* and just as quickly lets out a cry of "*terror, I've never heard a cat scream like that before.*" Another cat, who had been injured by wooden planks being tossed at him and, "who had begun to answer to my calls, suddenly became extremely wild like before."[12]

Humans create, accentuate, and justify the fear shared among the cats, and *this fear makes the humans laugh, giving them pleasure, reinforcing their contempt*. Such codetermination serves as a support for a feline psychology that the individual transmits, perhaps by way of a preventative genetics, confirmed through the education of the young, and the young then imitate the parents or those nearby through such encounters, the whole structuring a way of being, an ecological culture adapted to *this human environment*. And yet, the cat from the construction site, which becomes friendly and familiarizes itself with the environment only to become wild once again, suggests that nothing is fixed: *a change in the environment* can lead to a change in the animal in an ongoing dynamic synergy.

Léautaud himself depicts and embodies the changing times. Having discovered the cats living in the abandoned bathhouse, he begins to distribute to them "nine cents' worth of liver and three cents worth of milk" the very next day. As for the cats living at the construction site by the street, on several occasions he tosses them "pâté" over the fence or slips them some liver under the door and even brings them warm milk since it's cold outside. For the family of cats living at the construction site, a covered alcove is created thanks to the help of several other compassionate humans nearby. For Léautaud is not the only person to bring the cats some meat or even provide "rodents" for the cats to eat: a streetsweeper, a maid, a concierge, other passersby, grandes dames of the petit bourgeoisie, militant animal rights activists, the wives of painters and a university professor, even a "concierge," even "everyone" residing all around them! And the writer cites other people he encounters here and there who take the initiative to look after feeding and caring for the cats in other areas of the city: a tiny woman with her own financial means, a wine merchant, a flower merchant, etc.; here as elsewhere, it's mostly women who seek to provide aid, "these charitable good women who care for the cats." Léautaud makes a masculine exception every now and then, noting several instances of charity on the part of a concierge or a roadworker who oversees his construction site, or several others who make claim to their . . . passive . . . goodwill toward the cats; doubtless due to the fact that the men exalt a certain kind of virility and don't leave a place for such charitable affection, at least in public as regards such cats living in the streets, for which Léautaud, willingly associable, perhaps even misanthropic, sees no cure.[13]

> All of this culminates in the development of practices for the protection of animals in the second half of the nineteenth century, thanks to the instigation of the British example in the creation of the Society for the Protection of Animals in 1845, and in 1850, the passage of a law that would punish anyone who inflicted public violence toward domesticated animals, and the creation of other associations by the end of the century. This latter period will see a social diffusion of protections within bourgeois society and within the modest working class of the city, where we also see the organization of shelters of assistance created by women who become more fo-

cused in engaging in daily life in the streets, as we see here. Another additional factor at play is the increasing enrichment and population growth in the city during the nineteenth century, which leads to a vulgarization and increase in the consumption of meat, particularly in the cities. Hence an increase in the number of butcher shops in each neighborhood and the subsequent possibility of offering cats the scraps from dinner, or even the possibility of buying small bits of meat or scraps such as beef lung, often given to the customers for free by the butcher, a practice evoked by other writers from this period, from Loti to Céline.

Such a human prodigality is also the result of the difficulty that street cats have in feeding themselves, as a result of the installation of trash cans and the collection of rubbish that we see beginning during this period, even if such changes are relatively slow and only really begin in 1750 with the massive arrival of grey rats, referred to as the wharf rat, from central Asia, knowledge of which would still remain unknown to our inhabitants of the Parisian printer's workshop in 1740. Bigger, stronger, and more aggressive than black rats, they quickly supplant them and relegate them to attics, taking over the basements, wine cellars, sewers, and roads thanks to their preference for humidity. With their arrival, most street cats are at a large disadvantage in the face of these big grey rats, who are numerous and combative. In one of the places visited by Léautaud, he notices that a one-year-old feline has "a tail that has been completely demolished, more than likely as a result of struggles with gigantic rats whose presence filled the location." Cats must seek out other smaller rodents or birds as food, otherwise they can turn toward the alimentary offerings elicited thanks to their increasingly thinning statures, which is often mentioned in documentation, leading people to now pity them.

In a complete turn of events, beginning in the nineteenth century, the shared feeling concerning the uselessness of cats in countering rodents becomes such that a number of people begin to seek out rat-hunting dogs, which are more efficient yet less numerous, and finally begin making use of apparatuses with poison. All of this only reinforces the poor image of street cats as well as the contemptuous violence toward them. But unlike in the eighteenth century where such violence seemed miniscule and not worth discussion, in the second half of the nineteenth century such violence leads to heated debates between humans.[14] Yet, according to Léautaud, indifference is still the reigning banal attitude. And he doesn't refrain from documenting it.

Léautaud harshly condemns two men who have made a game out of cornering a frightened cat. "I repeat that I find it shameful and call them brutes and pigs." He seeks out the police and references the new animal protection law when workers throw bits of plaster at a cat as well as himself. He stands up to a concierge who reprimands him for tossing a bit of pâté into a small enclosure where the cats reside. As for the cats themselves, he hears brief, repetitive sounds (Léautaud calls out to them, speaks to them, more than likely making use of words) with their hushed cries (high pitched voice), bearing more on listening than fear, and they see a tolerable mass (a lowered posture, less threatening, more reassuring), which, taken as a whole, function as a signal addressed to them. *The women must also act in the same way*, and the cats slowly respond, invited and attracted by these particular repeating sounds along with the scent of meat that they can locate at a distance and differentiate from the others. This additional food grants them an economy of energy and time as well as the possibility of doing something else besides the uncertain tracking of rodents: allowing them to sleep, to lazily sit out in the sunlight, to play, to partake in a leisurely walk, and to be on the lookout for taking in information from their environs.[15]

> The evocation of relations between humans and cats leads me to refer once again to the utility of the social sciences. To the extent that ethologists insist more and more on the diversity of individual attitudes or those experienced throughout each of their lives, they discover that one's education, temperament, and environment play just as important a role as that of age and sex—two biological aspects that had been given privilege for a long time—and that the relations between a cat and a human are dynamic. Ethologists then begin to engage in psychological, social, cultural, geographical, and ecological approaches. Obviously, this is done through their specific education and specific methods of observation, which leads them to privilege certain aspects, such as the factors of encounter more so than their modalities, and as far as the latter are concerned, focus on tactile, visible, and measurable exchanges judged important from the privileged viewpoint of humans rather than via olfactory modalities.[16]
>
> Their method still encourages us to take a candid look at the social sciences [*sciences humaines*], which tend to think the plasticity and complexity of relations among the living. In order to study the relations of cats among themselves or with humans, we must enrichen our analyses through the sociological and anthropological study of interactions (symbolic inter-

> actionism), of the processes of the elaboration of knowledge and behaviors (ethnomethodology), the distancing of other bodies (proxemics), repeated modalities (rites of interactions), the emission of signs (semiotics), etc.[17] We must construct a sociology, a geography, an ethnology of the cat behaviors. Taking up this innovative task, two sociologists have initiated such a method and demonstrated that the groups, whether feline or humano-feline, must be analyzed in terms of sociology and that the domestic cat is a social being.[18]

In the early days, in what amounts to an atypical encounter for them, the cats remain vigilant and anxious. They more than likely emit signs (postures, odors, actions) reflecting and signaling these emotions, with an intensity proportional to the specific personality of each of them, to their repertoire of expressions, specific to each of their singular life histories and the unusual nature of the encounter. Such as in the case of the cat residing in the construction site alongside the street, who immediately flees *the first day Léautaud gestures toward him*: —ear-turn, quiver-sniff, pupil . . . sound, scent, mass: danger . . . scatter . . . "bones protruding" lurk and sneak, flatten . . . wait, paws on the ground, observe . . . attractive aroma (pâté is tossed) empty stomach effluviate repelling [human] . . . wait paws on the ground ears at the ready . . . scent of danger weakens . . . disappeared . . . keep paws on the ground . . . reassuring nocturn . . . go out . . . eat . . . —

> Surely some among you think that I'm trying to make animals speak, that it's useless to do so since it would only spill over into a banal anthropomorphism. And you'd be correct if that's what I was attempting or if I were offering you some kind of transcription of a stream of consciousness. For that would be to deny the originality of the species and individuals, to refer everything back humankind through a conjoined anthropocentrism. And it would be paradoxical to want to take the side of animals while simultaneously bringing them back to the human. This is not what I'm attempting, nor is it a question of deciphering and then translating some sort of animal language, in this case a feline language. Such a chimera is impossible for humans, who are not absolute spirits, merely members of one animal species among many others, enclosed within their biological and cognitive framework.

Rather, I'm attempting to respond to a question: how can one write in a creative or poetic manner so as to reproduce the animal viewpoint or perspective, so as to reproduce a feline individuality, to help you to grasp it, to feel and experience another form of lived experience or way of feeling other than our own? One shouldn't merely consider such an attempt as futile and unscientific. Even if such a poetic scientific process has been poorly viewed since the advent of the great division between the arts and the sciences, a number of scholars today still believe we should nevertheless attempt to return to a way of doing research that is both poetic and scientific, not in order to say whatever nonsense that comes to mind but so we can improve the way in which we express ourselves, our research and our ability to produce scientific discourse. For every scientific discourse is also a literature, a mise-en-scène, a performance, even in ethology and, of course, in the writing of history. The question is therefore all the more vital and the use of poetic language shouldn't be quickly discounted in regard to scientific practice.[19] In effect, it is possible to use words and even poetic literary writing as a kind of supplementary scientific instrument, but not in order to invent imaginary things, since everything presented or expressed is based off actual verified documents: everything discussed has been deduced through indications in the literature from which we are able to make inferences. Everything presented or expressed thus intersects with contemporary forms of knowledge in such a way as to perform or stage reliable data, but it is done so in order to help us rediscover and evoke forgotten situations of nonhuman lives from the past.

As I've done in *Animal Biographies*, in the following pages I seek to play with two facets of writing. The first connects us to humanity, since our language is the imposed means for expressing in human and be understood by humans. There is no other way. We are obliged to transpose animal reality, here that of cats, within a human framework, and to do so is to have to twist and contort such a reality while simultaneously employing a mathematical expression that some believe to be more objective and held at a proper distance. However, the second facet can grant us a modicum of freedom. Writing is supple enough to allow for processes (in the most noble sense of the term) that disturb or disrupt the reading, inviting or inciting, helping the author or reader to become decentered

> and leave their habitual world for a moment. Such a writing is merely a human artifice and not an impossible transcription of some nature. And yet, if one adheres to the characteristics and traits of the species and individuals in question, shining a spotlight on them, this can help us to project onto the animal side, attracting attention to events that have been neglected by humans, to perceive more broadly: grasping, experiencing, helping to understand perceptions and feelings, lived experiences that are indeed quite different.[20]

Slowly but surely, the cat emerges from his shelter a bit more quickly, as soon as he gets a whiff of the scent of food, but only "*on the condition that I keep my distance*, almost to the point that the cat can barely see me." The cat only accepts food as a mode of interaction while retaining a required distance, which the writer, through empathy, understands and respects. Léautaud adopts the feline's viewpoint while the latter perceives this sympathy and reacts accordingly. All of this more than likely took place through an exchange of glances, postures, scents, perhaps pheromones (but this aspect is still poorly understood as far as humans are concerned, perhaps even involving the eventual detection of exo-pheromones between the two species), or even through sounds, even though this cat doesn't seem to meow upon the arrival of the human or purr upon being satisfied, which *the writer would have noted as he did for the other cats*. Little by little, the feline lets *the human get closer in order to hand over his gift*, modifying his behavior in changing his perception/representation of humankind. But he immediately flees as soon as he has finished eating if the writer is too pressing, having felt he had convinced the cat of his benevolence, poorly judging this exchange, perhaps even taking to returning to his enclosed refuge, swallowing scraps slipped under the door "without putting himself on display, taking each bite while regaining his distance just as quickly." In this way, he shows how he adjusts his behavior with prudence and requires some distance. Only with regularity and more time will such stray cats become more serene, like those cats living in the Jardin des bains, *who come out at 4 a.m. as soon as the streetsweeper/nursemaid arrives* but who remain behind the gates and fence, maintaining their distance as well.[21]

They therefore don't rub up against these humans or deposit their scent on them; neither do they integrate them into their world, which remains composed of objects and places, along with a territory within which they merge. This is not a result of some natural atavism on the part of cats, inherited from their wild ancestors, for this book will show that they can certainly behave otherwise; rather, it's a result of the cats' interacting with *the same environment*, on a daily basis, which reinforces their being. Through a partial contempt for human intentions and through inattention, having slowly but surely accepted *Léautaud's presence in order to allow him to leave pieces of meat for them*, the cat living in the cobbled enclosure in the construction site ends up being caught and placed into another area *the writer deems to be safer*. As soon as the cat is freed, he quickly runs into a hangar and doesn't reappear for dinner, despite being called and sensing the smell of food, dreading being captured again. And as soon as night falls, he flees this new place that has yet to be marked and remains unknown, returning to his shelter at the construction site, to his wooden skeletal enclosure. The cat will only come out again the following day at the first sounds of a worried Léautaud, perhaps attracted to other signs *that this human emits without even noticing and therefore does not mention in his journal.*

The cats hold fast to their territories even in times of their very destruction. In spite of his fears, the same feline remains at the site *during the very construction maneuvers to knock it down, as workers begin carrying off old materials*: "The beams beneath which the cat resided had begun to be removed, placed into various piles, without however creating any new sort of enclosure, and so the cat sat there, attempting to hide under the very last beam, with a look of being uprooted, lost at the foot of the shelter which had now been destroyed." *These words are not merely the human projection and invention of a feline state; rather, they are the translation into a human language of an emotional reality that was well perceived by the writer*. The cat withdraws into the last vestiges of the enclosure that are still intact, then moves behind the latticework, and finally beneath the remaining beams. Cornered, he finds himself caught *by the construction workers*, which provokes him to flee as soon as he is released.

Léautaud searches for him in vain throughout the other buildings on the construction site (one enclosure being 200 meters and the other 150 meters) and catches a glimpse of him several days later at the back of a courtyard of a building located nearby. The initial terrain that the cat had refused was only 250 meters away. Such a distance is considered close for humans. But for these city cats such a distance entails an entirely new world with different sounds and scents, obliged as they are—by their number and density—to reduce the expanse of their vital living area. However, this particular feline must have marked the contours and periphery of his former enclosure and therefore probably had already lurked between the gate and fence protecting the courtyard from the street. He rejoins "*several other cats without owners, [. . .] who live as they please: a wine cellar remaining open to them and at their disposal day and night. All the kitchens in the building provide them with something to eat.*" Food, protection thanks to surrounding fencing, refuge in the wine cellar, *an empathic tolerance from the couple who are the concierges* must have incited the cats to stay, to mark the territory, to tolerate the others, to correspond and regroup their demeanors, mixing their scents together, weaving social bonds through signs and postures, even integrating *this couple of concierges* since some of the cats brush up *against them when they discuss with Léautaud*, perhaps because they are not stray cats but cats who've been abandoned or are vagabonds, moving from one house to the next. Such behavior is not enacted by Léautaud's "wild" cat, who nevertheless responds to the *writer's call* in order to partake in a food offering. Through attending to such adjustments, these streets cats are able to maintain their situation and even live fairly comfortably *as the writer himself notes in regard to the felines living in the Jardin des bains, fed everyday by one or another person.*[22]

And yet the fact remains (and this is another novelty of the era, another environmental change for the cats) that Léautaud, and the militant bourgeois women that he rubs shoulders with, are tormented by the risks cats run by remaining strays: cold, hunger, impatience with humans who only tolerate them for a short time, and the easy encounters with violence on the streets. As with others who seek to help protect animals, the writer will go from street to street, door to door, seeking to place the cats he is able to catch in the hands of people he implores and solicits: artisans, merchants, concierges, renters who live on the first floor. The results are mixed, with many people refusing to house a cat, others who quickly accept through a series of informal or almost nonexistent exchanges. A second solution arises: he seeks out the shelters created for cats and dogs at the end of the nineteenth century in many large Western cities, including Paris, run by landlords, often members of the SPA in France and often women, but he fears deficient hygiene, contagious promiscuity, and even the eradication of cats by poison gas, which had been in use since 1903, and he'd much rather simply leave them free and independent.

And besides, stray cats often don't appreciate such *placements*, for fear of the unknown (a new place where they have no orientations and that is not their territory, fear of humans who get too close, submerging their world through scents, sounds, and gestures), which hinders its very decoding. As in the case of the male kitten from the construction site who, after being placed with *a good caregiver*, continually seeks to flee, even trying to jump out the window. Or the kitten left in the courtyard by his mother and who Léautaud ends up taking in himself, a kitten who experiences four days of nothing but fear due to being uprooted. The unsettled kitten finally calms down thanks to the presence of other friendly dogs and cats, most notably a female cat who adopts him and lets him feed on her milk, but he quickly dies due to a virus, as he was more than likely not immunized, in contrast with the others.[23] In fact, it's the cats . . .

Abandoned, Suspicious
(Paris, 1908–1912)

. . . who appreciate such actions, with prudence.

Léautaud discovers such cats again and again near his work, in the Luxembourg Gardens, which he regularly visits in order to help them since the area serves as a cat drop-off for locals who want to get rid of them and who believe the cats will be able to take care of themselves in the park. Other abandoned enclosed areas serve the same purpose, such as the old coal factory where one day our writer discovers "a new, beautiful, strong cat who seems to be healthy" who has either been cast aside or has run away and is lost. Such cats live within fluctuating circumstances. Former house cats turned wandering vagabonds could return to being sedentary if they *were gathered up* or veer toward being definitive strays through an increasing line of descendants who have never known human socialization and acquire a distance from it. And once again we see evidence of behaviors that can be transmitted epigenetically and reinforced in each successive generation through an initial apprenticeship and daily experience. The cat from the coal factory represents such a possibility since it's an Angora: *a breed of cat once held to be rare and precious, reserved for a small few, but which became popular and widespread* thanks to an uncontrolled reproduction of the breed to the point that it has become relegated among the other breeds and is hardly doing any better: moreover, the wild cat living at the construction site is also "a beautiful Angora," a descendent of an indoor cat [*chat de salon*]![1]

These felines reconstruct their territory. According to Léautaud, who locates and follows them, the cats from the gardens adopt a rather circumscribed area and eventually change their location from time to time when the pressure exerted by humans or their peers becomes too strong, similar to what we still see taking place today. Some cats migrate. Some of them return and remain close to their original environment, *such as a cat the writer placed with a hairstylist near the rue Assas* but who is quickly left at the Luxembourg Gardens only to take up residence in an area quite close to his ex-owner. Other cats seek out a new domicile by themselves or by way of a peer who attracts them. Such is the case for an old cat from the park, *whom Léautaud places with a concierge*, and who regularly returns to the adjacent garden, even returning there one night with three other female cats who take up residence in a shed, one of whom as well also returns to the park each night. By necessity, these felines seem to have become more territorially flexible and take quick to their markings.

> I'd like to stop here for a moment and comment on this idea of adapting to the world, since it refers back to two conceptions of knowledge implying that biological and cultural approaches can bear strong differences that are difficult to reconcile even if such reconciliation must be attempted. The first approach, which was forged and propagated throughout the twentieth century by biologists and then philosophers, claims that each animal species constructs its representation of the environment in terms of its perceptive capacities and subsequently lives not within a common environment but within its specific subjective and particular world, different from that of other species.[2] This version takes an interest in emphasizing specificities but neglects relations between species, which would be considered difficult if not impossible if we admitted them as such. The other, more recent conception, developed by psychologists and anthropologists, maintains that, on the contrary, beyond differences in perception, the environment is common to all species; it invites them to interact, among themselves and individuals; the environment therefore plays a role in the confection of every individual being, its psychology, mode of action, and knowledge [*connaissance*].[3] In fact, biologists often consider that individual beings are constituted very early on in their development (with a very early socialization that sets into motion the biological machine) and that they then make choices and adopt behaviors without changing. Sociologists and anthropologists think beings in terms of an endless (re)construction through their relation with the world, but, either neglecting biology or setting themselves in opposition to it, they often forget that the worlds are specific to each species. This is largely because throughout the twentieth century, naturalists and philosophers emphasized the differences and airtight separation of these specific worlds, declaring that no two species would be capable of understanding each other. Officially this was done so as to avoid any anthropomorphism (which in reality is anthropocentrism) by rejecting the notion that any other animal could also be conscious or possess intelligence. And yet these same researchers ended up tossing out this scientific notion with the ideological bathwater.

> Both sides have forgotten that the existence of specific worlds does not prevent interactions, including all the comprehensions and misunderstandings that go along with them, without which domestications and the entire subsequent history of animal relations with humans would have been impossible. One must therefore believe in a relative community, a partial recovery of specific worlds where successful interactions take place based on scent, sounds, and colors. And one must retain this notion of a specific world while also forgoing its lone biological meaning, by complexifying it. The idea of co-construction makes one think that the specific world of a species can be adjusted within each era, for each group and for each individual according to different environments. In effect, sensorial capacities are implemented *and* more than likely modeled within each environment, similar to what we see take place regarding meowing and purring. The environment and the individual—of a place, group, or era—will therefore modulate or adjust into a specific world and will perhaps in time transmit this modulation through epigenetics. Such an optic of confectioning beings within the environment leads one to pay attention to "being-in-the-world," to its structuration in connection with this world (constituted with other beings and physical bodies), to its creation of a sensibility, of a psychology, of a knowledge, an action, of a culture at the level of its ecological dimensions, that is, its cultural dimensions in connection with the environment. And as the various beings will act upon this environment and influence it, they will transform it within a continuous and reciprocal process that will therefore not be a simple zone of contact between already constituted beings but a space of interpenetrations, of what we call a co-construction, a co-development of the being and its environment, within a co-synergy that would then be a historical process with, and giving rise to, beings and historical environments.[4] The historian can only be sensitive to this process. You will see that the idea of co-development—in this case between cat actors and human environments or vice versa, appears frequently throughout this present book.

Léautaud says or perhaps knows nothing about *feline sociability within such spaces and only provides several hints as to their rhythms of life*. They seem to scurry off and hide during the light hours (during the day) and install themselves under the odiferous shadows (*in the foliage of the gardens*) that are well located, inventoried, and known. They sit under shelters seeking refuge from the wind, rain, and cold (inside *the shed or under the hangar residing at the coal factory*), bereft of the scents, sounds, and dangerous movements (*humans*), and only come out at night (*in the evening*) to hunt or beg. We then have the arrival of the yellow cat into the territory, wandering down the neighboring street meowing at the risk of colliding with *projectiles cast by exasperated concierges and local inhabitants*.

The memory of having been more or less nourished and perhaps having acquired a few lessons in hunting—an essential retained skill—leads them to not be stressed or alarmed *upon the writer's approach*. Quite the contrary: the yellow cat from the construction site—ear-turns . . . *calls out from behind the walls* . . . *recognizes* . . . associating *human* food . . . ear raised tale high leaps then activating his paws . . . displaying them and holding them up high [on the wall] . . . quiver-sniff pupil catching a whiff of the drifting scent . . . pushes his paws "forward" placing his paws chews . . . begins again . . . —

> You must be asking yourself why I make use of such strange expressions: "push," "leap," "activating his paws," and not "gets up," "jumps," "runs." Another way of writing on the animal's side is to adopt a more concrete language, far too often neglected in the West since the advent of humanism, rationalism, and the scientific revolution, along with scholarly instruction, which have generalized abstract terms relegating feelings and gestures to the background while championing intellect. It is, however, possible to seek out and take a cue from more ancient languages and forms of writing, such as Akkadian, used in the writing of *The Epic of Gilgamesh*, edited in the second millennium BC. [In that text] we see the use of the expressions "feet carry you" instead of "get up," "turned his chest" instead of "give in," "Life in one's hands" for "save oneself," etc. So, in transposing on the side of cats, one can write "leap" or "place one's paws" rather than "jump" or "stop," etc. Rest assured such writing is not an indication of a lack intellect or a mere rudimentary capacity on the part of humans or cats. Rather, such writing allows for highlighting the profound entanglement between feeling, thinking, doing, the senses, cognition, and gestures that Western cultures have also demonstrated. In *La Chan-*

> *son de Roland* (eleventh century), the first mythic tale written in French, we can find expressions such as "the clear sun" for "a lovely afternoon," "sprang to one's feet" or "leapt to his feet" for "getting up" and "leaving," "stayeth his feet" for "putting on shoes," "weight" for "pain," etc. Reestablishing a bodily foundation to language allows one to value the bodily expressions of animals.[5]

The female cats from the shelter come running to *the entryway* and pounce on the pieces of food, prepared to depose of the most fragile morsel, no heavier than a "newspaper." Some of the cats in the park take on habits of place and upon the arrival of the feeding time, just like these cats "soaked to the bone, gathered in front of *the kiosk de la porte Fleurus*, wait for their pâté" as the hours of light come to an end *around 7 p.m.* In this way, they demonstrate their habituation to interactions with humans, showing they know how to decipher *the latter* and dialogue and meow so as to echo their calls, demonstrating they have lived within communities of the world. And yet, if these cats don't demand for a space to remain between themselves and the humans, approaching and letting the humans come close, most of them will still refrain from rubbing against the *gift giver*, from marking them, from integrating them into their new world, save for one cat who lives at the coal factory: "so familiar, so loving, following me in the street," but only in so far as the writer does not lead or take him. These cats have no doubt experienced *human harshness*, distant relations, and obviously life events that have caused them to latch on to a space and carve out their territory, their world.[6]

> You see that in adopting the social sciences and their idea of co-construction, one must become attentive to several different aspects. Firstly, to the perception of these cats, to their sensations and emotions, their sensibility and character, to the organization of their relations with the physical environment and the other animals including humans, to the expressions created by such relations (exchange of postures, signs, gestures, movements, gazes, sounds), to the interpretations used, to the various attitudes and knowledges acquired. So, one must become interested in the construction of their culture, envisioned here as a process of acquisition for being and surviving in the world; as an acquisition of forms of knowledge or know-how: knowing-how-to-perceive, knowing-how-

> to-do-act-react, knowing-how-to-be. And finally, one must think of the endless construction of behaviors, relations, and societies, from the point of view (define the situation, choose, act) and from the world of each being, from this very being itself, from his self and his or her gaze (in the broad sense), always in a provisional state as the social and the cultural constitute it, in scrutinizing the adjustments and changes. But here again, all this would do nothing more than seduce the historian: the apostle of dynamics.

Léautaud is encouraged by the good contact he has established with various cats in order to capture them. But he is also on the receiving end of recriminations from concierges who don't like him feeding the cats and who want to ship them off to the Luxembourg Gardens. As a result of this, Léautaud captures and constantly places them with the same kind of owners as the stray cats. And some of his attempts are successful, such as in the case of the little kitten sitting quietly purring on the bed of a welcoming caretaker and who just as quickly falls asleep. But there are also *many failures,* such as the feline *gifted to a hairstylist* who is hardly fed at all and who is quickly dropped off at the garden due to "eating a bowl of cheese," or another, "a poor schlep" placed several times with different people and *finally handed over to a concierge who puts him in the wine cellar "with his bed, his bowl of milk, and his food," no doubt in order to send the rodents packing without ever truly thinking of creating a more personal attachment to him.*[7] *Moreover, Léautaud never demands or requires this of the caretakers, merely hoping for the gift of room and board on the part of the human;* the gift of attachment is left up to the cat.

Given such a context, these cats who are constantly displaced hold fast to their emotions and only worry about their territory, while also making themselves just as flexible as stray cats to overcome their challenges, only integrating the human into the margins of their territory, as an energy support, thus also saving the humans time and effort. Such a manner of being constitutes an adaptation to the *human environment*: in this case modest city dwellers oscillating between a disinterest established for centuries and a newfound interest, driven and proposed by the bourgeois, of ownership, perhaps even a connection; an often vacillating, ephemeral desire giving way to a fragility of adaptations and quick abandonments: a prompt refusal of new bourgeois attachment and a return to a more popular principle, to a more hardened conviction that cats should fend for themselves.

> No doubt, you must be saying to yourself: well, these cats—*our cats, your cats* are good at adjusting. Obviously. Ethologists have finally begun to accept this claim, which now somewhat justifies rolling out approaches heralding from the social sciences, not so as to replace others but as a complement to the natural sciences. Such an approach favors taking into account all the dimensions that are at work, from all the explicative angles (and not "levels," which would suppose some kind of successive hierarchy whereas everything plays itself out at the same time through the physiological, the cognitive, the psychological, the ethological, and the sociological). Such an approach must lead to the mobilization of the so-called nature disciplines, which would be more aptly referred to as biological disciplines, as well as those referred to as the human sciences, which should be renamed as cultural and social sciences, and applied to animals in such a way as to demonstrate that cats are biological, cognitive, social, cultural, and historical beings, like so many other species as we are continuingly to discover more and more.
>
> [The above is desirable] on the condition of refraining from the temptations at the edge of each discipline. Placing oneself between humans and animals, the zoo-socio-anthropologists have a tendency to eliminate biological differences in order to only see the interactive common gestures. But, if we really want to stand on the side of animals, we cannot forget the biological and its sciences.[8] And yet, these same researchers often refuse to take into account ethology, which they accuse of biological reductionism, and prefer instead to focus on the social and cultural sciences and their methods. Such a withdrawal into their discipline mirrors that of biologists. The latter only think and believe in terms of evolutionary, genetic, physiological, neurological, ethological, and ecological explanations and therefore only see the cat as a biological being. In so doing, they construct other impasses. Officially such impasses are constructed for fear of anthropomorphism, often under the influence of anthropocentrism, which leads to a refusal of attributing human capacities to other animals in order to preserve humanity's supposed privileged abilities. One must therefore take a step back from the temptation found

on both the cultural-social and biological fronts attempting to explain the entirety of the situation: a social whole, a cultural whole, etc.

One must also steer clear of the danger found in certain research by a rare number of socioanthropologists who focus on humano-animal relations.[9] Their idea of a constitution of self through interactions leads them to veer toward a proximity of beings and their capacities when such relations are achieved. It's certainly true studies have shown that cats and humans share homologous cerebral structures and functioning for social behavior, arising out of a shared ancestor. This allows for interactions and socialization between them, with similar neurological and physiological mechanisms within the expression of behaviors and emotions. However, proximity does not mean identity. One shouldn't neglect the specificity of each species. In fact, those expressions made by cats, often forgotten or aligned with those of humans through an interactionism, detached from the biological and the cognitive, have led to a rampant humanization. This is because their relations always activate capacities and dimensions that are not shared between species, often not perceived or poorly perceived by the humans, such as smell, scent, hearing, and sounds that are all very important for cats but remain of lesser importance for humans. Their interactions therefore partly convey a reciprocal incomprehension. And this incomprehension creates wobbly, unstable, and complex adjustments. The cultural disciplines must therefore take to attending to both the biological and the cognitive.

In complete contrast, the biological disciplines must loosen the yoke they have imposed over animals: an evolutionist and genetic governance that is rigid and direct, since this has led to the belief in unavoidable and identical behaviors for everyone. This still leads to research on contemporary domesticated cats through looking back at their wild ancestors or cousins, as biologists maintain that the first domestication hardly had any impact (since the genomes for wild and domesticated cats are very similar whereas the morphological differences are few) and that little has changed since earlier times.[10] This is true to such an extent that zoologists have recently suggested that the domestic cat is not a species but rather a subspecies of the wild cat and renamed it as *Felis sylvestris catus*.

But such a claim merely accounts for a single dimension of the living. Such claims still rest on a genetics imposed over all sorts of behavior that was promulgated by geneticists from the 1970s through the 1990s. However, contemporary epi-paleo-geneticists no longer adhere to such a mechanical vision and on the contrary point to the role the environment plays in genetic expression, whereas ethologists reiterate that a theory of proximity does not forbid behavioral differences between wild and domestic cats. Furthermore, the biological prism prevented us for a long time now (or still continues to prevent us) from thinking about the difference between individuals or between groups within space. It also leads us to be unaware of the oscillations within time, be these in the short or long term and which this present book evokes, which clearly function as adaptive respirations.

These latter adaptive respirations—and this is the entire reason for the history signaling and suggesting what will follow—wait for other sciences to validate them; they must rely on other dimensions than genetics. Even though it begins and still focuses its investigations on the biological, new research in epigenetics shows that gene expression is not automatic.[11] It is facilitated or hindered, even prevented thanks to the environment. The resulting epigenome—like a musical partition within a much larger available genetic piano keyboard—is variable among individuals within the same environment and obviously within another environment once individuals change within space and time. Even if behaviors are hardly dealt with in epigenetics, several facts suggest that the environment could also play a role in their regard. Following this line of thinking, there has been a study proving that when a mother rat licked her young in a certain way, it incited in the rat infant the methylation of a gene producing a protein reducing the stress hormone, which subsequently conferred on him a calm temperament.

Such studies only reinforce the recent idea of a behavioral variability within space and that of a flexibility within time (which will be discussed in the middle of this book). The eventuality of an adaptive transmissible and reversible epigenome would also encourage the hypothesis concerning intermediary historical times between individual time and the time of evolution, times flowing for several generations in which genetic expression will oscillate without the genetic capital changing, through inciting the oscillation of behaviors. The idea that epigenetics allows for the acquisition of ways of being that are temporarily adaptive during several generations is hotly debated in the realm of behaviors whereas the idea of an inherited epigenetic capital is more or less fought against. If there is a certainty regarding the fact that the environment modulates genetic capital and that epigenetics is one of the intermediaries allowing for this, we still can't affirm such modifications are inherited at the scale of individual generations. They are certainly inherited at the cellular level, but the debate remains concerning the capacity of these modifications to pass through the germinal lineage and be transmitted from parent to child. This doesn't mean that such transmission does not in fact occur, but it is contested.

And yet epigenetics still maintains the connection between the biological and cultural sciences. Since its thesis of a modulated innateness suggests a functional connection between the innate and the acquired, one must tie together the disciplines in question. Epigenetics also shows that the environment must be considered in a larger sense: not merely in its physical dimension but also in the dimension of the surrounding beings (the other peers for a domestic cat, or even the various surrounding people whose attitudes weigh heavily upon it) as well the structural dimensions (be they social, cultural, economic . . .) borne by all involved, all the things that can be studied from both sides of the sociocultural and biological disciplines.

These varied disciplines and dimensions lead us to believe that behavior is not merely the fruit of some kind of biological impulse or drive or a genetic determinism but is a result of a bioenvironmental adjustment starting from a store or conglomeration of adaptable capacities in relation to an endlessly changing environment—behaviors where the biological and environmental dimensions (physical, social, cultural, of the species in question along with the neighboring species) become intertwined, compounded, and entangled. It's therefore plausible to think the biological is in need of the environmental in order to become expressed, so as to avoid disordered manifestation and to become embodied as adapted biocultural behaviors. No doubt, it's with such biocultural behaviors that . . .

. . . these cats adapt, connecting little by little *with humans*, reinforcing *the certainties of the latter*, and on and on. However, street cats can adjust themselves in a different manner, within another space, *between streets, gardens, parks, city dwellings, or village homes*, and become . . .

CHAPTER 2

House Cats and Garden Cats

. . . since they adopt a domicile, transforming it into a central hub of their territory while continuing to come and go as they please. It would seem that garden cats are rarely stray cats (which are too distant); rather, they mostly comprise cats who have been abandoned or are vagabonds moving from one residence to the next, *compelled to the next abode by way of another context, a new environment that emerged starting in the nineteenth century: [the influence of] humans, often bourgeois but also people from more modest means, who tolerate them not for their utility at hunting mice—for which many others appreciate them and which was the most common case in the past—but out of pleasure and enjoyment.*

> For they no longer partook in the old representation of the alley cat. The reappreciation of the cat had already begun to emerge in the preceding century: one of the first individuals to speak in favor of cats was Moncrif, who exclaimed that "cats were admirable examples of activity, modesty, meriting a noble emulation, and seemed to detest laziness" and that the black cat was the "most beautiful color of them all." This tendency will become reinforced in the second half of the nineteenth century by artists such as Manet with his famous portrait of two alley cats, one white and one black, thus creating an equality between these two colors and a certain conception of contrasting symbolism, and writers such as Banville,

whose admiration went all the way to considering them as the most aristocratic of animals ("both aristocrats and cats partake in the same level of nobility, a certain self-respect, and elegance"), and Mallarmé, whose praise for the cat elevates it even higher to the level of an archetype of "cats, lords of the rooftops."
Such a reversal regarding human appreciation of cats comes from the Romantics, who had also elaborated a new representation of writers and artists: misunderstood, even cursed, solitary and poor, yet autonomous and free from mundane patronage. Beginning around this time, many a writer and artist will consider the cat—in particular, the alley cat—as a kind of alter ego and subsequently attribute a similar character to it: wild because it is free and independent; suspicious, disobedient, and even righteously ungrateful. The Romantics' reading of cats is reinforced by the rediscovery of medieval witchcraft, which validated the notion of the mysterious and maleficent cat and provoked them to praise this other marginal creature as an image of themselves. Such a process of a double myth of the artist and the cat—with the projection of one on the other—can also be seen playing out in Anglo-Saxon countries, where this reconfigured beast populates fantasy literature as well as the detective novel, beginning with Edgar Allan Poe's 1843 work, *The Black Cat.*[1]

As such, the great writers of the day leave glimpses in their writings of several street cats who have come to seek refuge in their humble abode, often coming to their defense, such as Edmond Goncourt, who writes that he "never really liked cats" but nevertheless gathers together a crew of them between his legs in his garden. Or Anatole France, who spots a cat who enters into his home on his own accord, "without any worry of being noticed," doing as he pleases, ignoring any commands, and who will disappear for eight days at a time before regularly returning, *as long as the writer requires no sort of obedience, nor imposes any kind of duty, appreciating as well this reciprocal freedom, and subsequently shares in this new representation of the alley cat, therefore creating a new environment allowing for the* expression of a new feline way of being. For this cat, christened Pascal, in reality is no longer exactly a street cat, since he has integrated a house and a human into his world and has adapted his way of being. Such a modification is more easily noticed in such a cat, "an alley cat," who fell from a tree in the garden of *Emile Goudeau,* (1849–1906), cofounder with Rodolphe Salis of the satirical review . . . "Le Chat noir" (1882), who created, in 1881, the famous Montmartre cabaret bearing the same name. Bestowed with a warm welcome, no doubt through signals such a generous amounts of food and speech directly addressed to him since he was called Mouchi-Moucha, the cat will hang around on the condition that "he is left to roam freely, in complete liberty," which perfectly suits the *bohemian spirit of this gentleman*. However, the very same feline quickly reduces the distance and ends up staying in the various rooms in the house, approaching humans, and eventually ends up modifying his own ways of being: he masters his own ways of creating a safe environment by climbing up onto the lap of the man sitting in the armchair, *who willingly accepts his company*, in order to thereby profit from the warmth found there and sleeps, but also feels his way, in order to nourish his perception of *this human*, rubbing, marking objects and beings, mixing their scents with his own, integrating them to his world but without entirely integrating them: he doesn't follow him and disappears once *Goudeau moves*.[2]

A more detailed description of this behavior is provided to us . . .

The Jaunets: Somewhere between the Street and Home

(Avignon-Orange-Sérignan, circa 1860–1879)

. . . by Jean-Henri Fabre (1823–1915), a renowned zoologist of his time, a specialist in insects, reputed for his descriptions even though his interpretations of animal instincts have since been rejected. We find his comments on cats in the middle of the second volume (1882) of his *Souvenirs entomologiques*, where he adds a chapter entitled "Histoires de mes chats," concerning the rotation of animals in order to disorient them, inefficient in the case of insects but practiced and suggested in the countryside for ridding oneself of cumbersome felines without having to kill them. It was such an ordinary practice that it forms the narrative structure for a popular children's song of the era, "Le Chat de la mère Michel." Placed into the dark abyss of a sack and swung around over and over again along the journey out into nature and then released, the cats would no longer be able to find their way back to the house. If Fabre admits that kittens can be disoriented, he objects to such a possibility in regard to adult cats by evoking "the series of yellows" [*sic*] he will give a home to in Avignon where he will teach at the lycée imperial from 1853–1870, and then at Orange, where he will dedicate himself to writing, and finally in Sérignan beginning in 1879.[1] This leads him to describe the attitudes of attachment of three males (the original progenitor and the subsequent two descendants) under very specific conditions.

The original cat shows up on a wall of the garden *one day at the beginning of the 1860s* and meows, more than likely quite loud in the direction of the inhabitants, thereby proving that he had learned to solicit *humans*, that he had already had contact *with them* and had integrated them into his world, and that he was either abandoned or a vagabond from another nearby residence rather than an independent stray, proving that he knew how to proclaim he was hungry while also betraying that he was a mediocre hunter and an often rejected beggar.

> Perhaps you think it's quite normal for this cat to meow to get food, as would be the case with your cat or your neighbor's cat? I'm afraid you'd be wrong. In studying current uses of the meow by cats, ethologists have indeed discerned that meowing has become fairly common in the West but remains quite rare between domestic cats, except for the mother with her children. Meowing can also happen during moments of conflict or relations and is also performed by the domestic

cats' more wild cousins. Ethologists therefore consider the increased use of meowing as an effect of domestication and interactions with humans, as a capacity that is present, henceforth intemporal, to be emitted by felines and to be heard by humans. The ethologists project their current observations into the past. They postulate a permanence in the act of listening to cats on the part of humans that quite simply did not exist. The example of the aforementioned cat demonstrates that his meow must have been heard by certain people but had also gone unnoticed by many more. The contemporary example of Toto, which will be discussed later on in this present work, even emphasizes an implicit refusal to hear the cat on the part of some humans. Humans of this earlier era were less predisposed to hearing a cat's meow than our current generation as a result of the differing representations and the way in which humans used cats in the past. Such indifference led Toto to be quiet, as was certainly the case with many other felines. Along with the variation of meowing in space revealed by ethologists between contemporary humano-feline couples, we also have the addition of variations within time. Rather than there being a permanent and consistent use of the meow by domestic cats, one must think of its use in terms of the human environment. So, one should posit here the idea of a (re) deployment of the cat's meow between the nineteenth century and our current twenty-first century, passing by way of, for example, Léautaud's Parisian cats who were already more inclined to meow thanks to the multiplication of the attentive human caretakers at their disposal, which hardly seem to exist within the more ancient landscape of Fabre's Avignon.

You can find all the reasons in the world to connect ethology (in the larger sense, which includes the biological, ecological, and cognitive sciences), ethnology (through associating anthropology, sociology, and geography), as well as history (in grouping together the sciences of the past) in order to build a diachronic ethology-ethnology, which I synthesize into a *historical eth(n)ology* of animals. How is it possible for a historian to build and navigate such a theory? He can only test out etholo-ethnological approaches, since these animals from centuries past no longer exist (for the most part) and the existing information concerning them cannot somehow be augmented. The historian must make use of various forms of knowledge in the midst of development in the biological and cognitive sciences in order to question and understand the historical documents at his or her disposal, or even read between the lines so as to make the proper deductions. The historian, however, cannot do the same thing with the cultural and social sciences since animals are hardly ever mentioned and are still not in a position to provide much information to us. But historians can nevertheless mobilize these other sciences modes and manners of analysis and reading in order to locate aspects that they wouldn't otherwise see and in so doing enrich their interpretation.

Tying together and articulating these diverse approaches, this historical eth(n)ology must focus on all the various aspects of the beings in question (bodies, gestures, interactions, interdependence) in order to approach the dynamic construction of their world, their ways of being, their know-how [*savoir-être*], their culture, in emphasizing the incessant and mutual influences within the environment, insisting on continual modifications in time and space, thus studying behaviors on several temporal scales: individual, generational, social, specific.[2] If the intelligence of this or that species and its individuals resides in the capacity to adapt oneself to the environment—as contemporary ethnologists claim—obligatory behaviors thus fluctuate, since the environment is endlessly in the midst of change as well. Therefore, cats cannot simply be studied within the contemporary Western lens of early twenty-first-century knowledge. They must be determined within a multiplicity of environments—both past and present while creating a history and geography of behaviors.

The children are the ones who respond to the cat's cries for food, not Fabre, even though he hears the cries of hunger and notes that the cat has "gaunt hindquarters and a bony back as a result of hunger," no doubt because he is inclined, like many of his contemporaries, to let the cat fend for itself. He excuses the children's "pity" by the fact that they are "very young," "giddy," and justifies his tolerance for a certain sentimentalism that he seems to judge, or wants to note, as something inappropriate, perhaps in the same manner others speak of cat ladies: tolerated or mocked due to their supposed belonging to the weaker sex. Justifications and criticism alike reveal the significant role that both women and children played in the progressive modification of human relations with cats up until today, and the distance kept by man, still noted by contemporary observers, even if the historian believes such distance to be not as severe today as it once was.[3]

Maintaining his distance through height, the cat stays high up on the wall. He has perhaps experienced being chased off or other forms of violence and thus this distance is vital. However, he willingly eats bread dipped in milk—the typical food ration for a cat of the day, meant to force them to chase their own meals *or leave the meat (less abundant in quantity at that time) to the humans.* But he grasps the treat on the "end of a branch," destined to give him everything he needs while still remaining a distance. The humans as well wanting to keep their distance, as cats often carried various skin diseases such as ringworm, which led to crusted plaque on the head, neck, or back, or other contagious infections such as rabies, which is fairly rare among felines *according to the veterinarians of the period, or mange, which was quite present at that time and for which there is no treatment. All these potential diseases and infections will dissuade a great many people from getting close to cats and accept them or pet newfound kittens. Such a trepidation on the part of humans only leads to more kittens being thrown out into the streets or killed, as Pierre Loti poetically describes in his work* "A Dying Cat" (1891).[4]

Jaunet, however, can eat to his heart's content among *this bourgeois family, where even though milk is less consumed by humans than it is today, there was still an ample amount so the family did not need to worry about rationing compared to more modest families of the day. Remaining suspicious, he still leaves in spite of "friendly" calls with soft tonalities circulating a detectable emotion* for these felines habituated to interactions. However, he reappears not too longer after, once again hungry, driven to return by his lack of luck elsewhere, recalling to mind his success in seeking food here from earlier days. After once again swallowing some milk-bread on the wall, he carefully listens for the "sweet words," no doubt adjusting his ears in order to record these transmitted emotions as well as signal his own, eventually creating a positive representations of these humans, since he climbs down from the wall and strolls in the garden, accepting to reduce the initial distance he had maintained and even lets *the children touch his back, who take pity on in his thin appearance and reinforce their emotions by reducing the volume of their exchanges even more in the same way that they might have hunched down to reduce their size in order not to scare him, therefore sending clear signals* for the cat to read since cats do the same thing. Jaunet can then gather up a sensible form of knowledge:— . . . "fur-vibrating" through touch . . . feeling, distinguishing, recording forces, harshness, manners . . . ear-turned to the sounds pupil to the forms quiver-sniff common [family] and different [individual] scents ROUNDED FLaTTEN head back . . . rub against the nearest person [legs or hands] . . . —quickly agreeing, having already lived through such an experience before.[5]

> You see there is a need to truly know our contemporary cats in order to understand, guess, and suppose their ancestors. But this comparison must be prudent. We must be careful not to link all behaviors back to the former, so as to avoid mechanically plastering their realities onto their ancestors', to preserve the particularities of their predecessors. Each time, one must start from the presupposition that the ancestors behaved in the same way. But it is also important to affirm that nothing is certain and be open to the fact that they may have behaved differently. In other words, one must avoid brutally administering etho-ethnological forms of knowledge onto historical evidence. If we did that, we would be generating an intemporal portrait of animals by way of contemporary representatives. To do so would be to deny their historicity when it should precisely be at the heart of historical eth(n)ology. History and eth(n)ology should reciprocally nourish, influence, and transform each other. It is therefore necessary to develop a historical dimension in eth(n)ology, to the extent that the eth(n)ology recognizes that the frequencies, intensities, and various combinations of modes of communication are actually multiple. Indeed, they depend on sex, age, initial stages of learning, the genetic modulation of each particular cat, as well as the social structure of a given place, between cats and humans, an aspect that obviously varies just as much within a human environment always in a state of flux. To such a degree that the conjugation of communication signals between colonies of cats appeared differently in those cats recently studied based on their individual social singularities.[6] This leads us to think that such was also the case in the past. Ethnologists eventually arrive at focusing on history but with great pains and already equipped with their fixed ideas. They claim that the domestic cat is nothing more than a tamed wild cat. Some today will finally admit that cat behavior has changed since its initial domestication, under the influence of social life surrounded by humans

> and feline groups. For example, they might agree that purring or raising one's tail high in the air are the result of ongoing change. But they do not know how; nor are they capable of retracing this history and have a tendency to think that all such changes took place during the process of domestication and that nothing has changed since.
> So, what we're asking is not to simply combine an uncertain science (that of history) with a definitive and complete form of knowledge such as ethology. Because ethology's portrait of cats also fluctuates. After having claimed the noncooperation of these felines, on the pretext that their wild cat ancestors were solitary and aggressive toward their peers-rivals, more recently ethologists have shown that good relationships between them are common. For a while now, ethologists have insisted on the utility and benefits of short-term collaboration for the proper socialization of cats among themselves or with humans. And they understand this behavior as a long-term survival adaptation.
> If it only leads toward an increasing series of endless interactions, an eth(n)ology-history should favor instability, which is the most fertile scientific situation; that is, a way in which each science is forced to reevaluate its questions, approaches, its acquired knowledge, and subsequently allowed to progress. Each of us must perturb the other, incite them to reposition, modify, and readjust themselves vis-à-vis their knowledge, whereby all of it functions as a work in progress. The eth(n)ological interpretation of a historical situation helps history to better understand, enrich, and renew itself. But such a practice also leads to questioning eth(n)ology and its flaws, forcing it to revise itself, which, once again, will lead history to be more vigilant. By encountering each other, the disciplines modify themselves.[7]

Elliptical in his further writings about Jaunet, Fabre above all emphasizes the role played by the children who, through their gestures, words, and doubtless their scent, food offerings, and the bedding they offer it, succeed in getting "the wild beast to stay" for an inexact amount of time. What does "stay" mean within the tripartite context of that time period, humans, and cats? The entomologist does not describe the feline's respite at his abode, reflecting more on his initial arrival and departure, but he does offer several hints about how we could characterize it. That he became a well-behaved "superb matou" (an expression often attributed to neutered cats; however, this wasn't the case for this cat) indicates that the cat transformed the house into a habitual site and principal location for provisions. Perhaps not the only site for provisions, since he retains, *and they grant to him*, his freedom to wander and roam, run off on "adventures around the neighborhood." Such a proximity implies that he must have shrunk the expanse of his territory since he no longer travels far to seek out food (like other cats in the countryside still have to do even today) and he must have remodeled his occupations, reduced his time tracking rodents or begging for food, and taken up the lazy leisure of observation, of feline sociability and mating. Perhaps it's even Jaunet himself who ends up bringing a female cat onto the premises, *a second cat which the family also discovers climbing along the wall of the garden and who attempts to get their attention; she as well appears to be domesticated*, thereby generating her litter on the premises and creating a lineage. By being in a better physical state of health, a male cat has a lesser chance of being subject to a number of illnesses, most notably numerous infections contracted from rodents and parasites, is less anxious and concerned with licking, scratching, and remaining cloistered, and is therefore more available to interact with his humans. *Fabre mentions two such instances*. Jaunet can rub against their legs, *since he accepts them*, first rubbing his head against them, then his legs and his rump, thereby mixing scents and creating a group, a friendship, marking and integrating them into his world and territory, *in the same manner the humans do for him, by petting him and letting him wander freely in the garden and the living room*. He also purrs around them, more than likely asking for more contact, for a caresses, or to express his satisfaction, his pleasure, all the things to which his humans pay attention and read in his interactions in speaking of "friendly purrs."[8]

> Are you still going to believe such interactions are intemporal, since we also experience such interactions today in the twenty-first century? Of course, cats purr among themselves on multiple occasions, most notably male kittens with their mothers or adult cats with other cats they are familiar with, and they learned to purr with humans as well during the process of domestication. And yet, there is nothing automatic about this latter possibility; its setting into motion depends on specific historical circumstances, and many other

> cats don't consider such purring with humans to be a positive activity. Stray cats, who maintain a distance from others, don't purr. And stray cats more than likely were the most populous type of cats in cities and villages all the way up until the twentieth century. Cats who have been abandoned to the wild don't purr unless they are reintegrated into a home or farm. Even in cases of reintegration, such as that of Toto, indicate that cats cannot purr if humans keep them at a distance, only keeping them around to hunt mice—which was often the only reason cats were tolerated in a great many areas. In order for purring to become a common trait of cats in relation to humans, the cats must already be inclined to approach humans and the humans must accept them, even draw their attention. Once again, only when humans take an interest in attending to the cat's needs will such interactions take place between humans and cats.

Instances of cats purring for humans must have been an interaction that took place in various time periods and locales, but the literature we have shows a crystallization of interest among the aristocracy of the ancien régime, and then a second wave of interest in the nineteenth century. In England, the terms "purr" and "purring" are coined in the seventeenth century. And such terms have their French equivalents, such as "faire son rouet"—literally to "spin one's wheel," an expression that a certain Théophile Gautier or a Pierre Loti still use, both of them writing, "spinning his wheel" or "to thread one's spinning wheel." This reference to a thread spinner implies that the activity is a kind of mechanical physiological response. As such, Moncrif makes mention of this tendency but is not sure as to its role and therefore doesn't give it a name. He does, however, note that it could be annoying: "a soft murmur, and which seems to be nothing more than a way to garner friendship." Following the social deployment, notably within the bourgeoisie, of the model of a pet cat throughout the nineteenth century, we finally see literature provide us with a concrete definition of purring: "the indication that a cat is satisfied and happy." French literature adopts the term "ronron" or "ronchonnement" to denote the dull continuous sound it will transpose onto cats (Balzac, 1824) and that term will lead to two others—"ronronner" (1852) and "ronronnment" (1863). The English don't create any new words but eventually bestow the same meaning onto the terms "to purr" and "purring." We could maintain that literature merely created several terms for an activity that had long since existed but was simply referred to in a different way. In reality, "ronronner"—to purr—has a much different signification than "to spin one's

> wheel," since it places the accent on the emotional and relational dimension of the activity, to the detriment of the physiological. It is therefore more credible to think that literature set out to innovate since the feline attitude is better listened to by men of letters more apt to grasp a certain nuance and express it. As such the very same Gautier will write "*ronron*" in italics in 1869, due to its novelty, and will then define it by way of an analogy to make sure he is understood ("a sort of cooing"), qualifying it soon after as denoting a cat to be "friendly and joyful."[9]

Does Jaunet experience other kinds of interactions: listening to the language of humans, meowing at them, playing with them? *Fabre doesn't speak a word about it, and it's perhaps noteworthy that he even writes a summary of their relation in order to justify moving the lineage of cats* with him, to "those poor beasts so often caressed." Furthermore, does the cat even hear his actual name—"Jaunet" (Yellow cat)—*which is the name the family gave him due to his tan color?* The text shows that the cat in fact is more accustomed to perceiving the name "Minet," *a term created in the eighteenth century to express affection contrary to the various contemporary uses and understanding of the term. At the time "Minet" was simply another generic word for "cat" and implies more a relation than a familial integration, equivalent to the name "Puss" in anglophone countries during the same era. And in fact, the name "Jaunet" or "yellow cat" is more likely used simply to distinguish the cat from others in the neighborhood (with whom he still continues to roam), in the same way the name "La grise" (grey cat) is used in reference to the cat cared for by Contat's printshop owners; later the name will be used to distinguish the lineage, each of whose members will also be named "Jaunet" and will then become "the Jaunet family." The term then becomes used not so much as a first name but a name used to denote a cat family parallel to humans without, however, being integrated with the latter. Furthermore, with each line written, Fabre indicates that he thinks of these cats in terms of a family or clan (just like Loti during the same period with his "Moumouttes" family), writing about the "old one," "the young one," and the "grandpa," and that he still situates himself within an era where one only refers to one's "cat" or "our cat" in cases where one must give it a first name to individualize it.*

When the Fabre family moves to Orange in 1870, they decide to bring the Jaunet family along with them instead of abandoning them, as many a family would have done, for it was not typical to travel with cats, since they felt it would be a "crime" to leave these cats to the misery of the streets and human violence. As such the Fabres impose a certain moral duty on themselves toward their cats, which was still very rare for that time period. However, only a couple of representatives of each lineage are kept: the most docile male and female kittens, along with the original yellow male cat. The remaining cats are placed with others. And on the day of their departure, the Fabres leave despite the absence of the grandpa cat, who had left to roam out in the fields. Nevertheless, they offer a reward to the mover if he is able to catch and bring the cat to their new residence, which he succeeds in doing several days later. This desire to bring their cat along with them demonstrates the Fabres still had a preference for the entire group of cats as well as a more specific attachment to the original yellow cat.[10]

No doubt the latter must have meowed when faced with an unexpected obstacle (the closed front door to the house), wanting to enter in order to eat, sleep. He probably called out again and again, meowing, until the obstacle disappeared (the door was eventually opened by the mover, who had completed his work) and quickly ran in, compelled by hunger, and hardly paid any attention to this unknown human stranger, instead being simply reassured by the permanent scent (of the living room) and the lingering scents (of his peers: his humans).—feels TRAPPED, CAPTURED CARRIED OFF . . . can't escape bristles HIS FUR LOWERS HIS EARs MOVES HIS PAWS ClAWS the AIR s'p'it'S MEoWs surPRISE ANGER finally Freed . . . —into nocturnal darkness (*into the covered trunk of a hippomobile*). He must have suddenly widened his pupils, leapt up and down trying to get out, scratching here and there to force his way out, disoriented by the various unknown scents, stressed by an enclosure that he had more than likely never before experienced, frozen, listening, meowing, trying to escape again, then waiting for a long time, spending *part of the afternoon and the evening inside.* —, suddenly, fragmented hammering sounds [the hoof beats of horses], breathing, cries [horses neighing] more unfamiliar cries [human speech]; a strong aroma in front of him, the feeling of cries raining down from above; shaking, jolting, rumbling sounds from below, hammering beats ahead, and then more endless aromas surrounding him; huddle close, then jolted, huddle close, then shaken again, leap, huddle close, fear, anger, stress of duration (*over eight hours to travel just thirty kilometers at a speed of four kilometers per hour*); noises cease, jolted, *cries* all around, a brusque ray of light, "de-pupillate," self-defense, leap . . . —"He came out of there a fearsome animal, with raised fur, eyes blood red, lips whitened with drool, scratching and breathing."

The cat is distressed *to such a degree that Fabre thinks he must be angry but then quickly realizes what has happened*: "This is the alarm of an animal who has been uprooted from his territory," *or what contemporary ethologists* refer to as territorial aggression. Jaunet doesn't flee, he doesn't try to return to Fabre's original land, it's too far away and he's too old, Fabre thinks, and perhaps no longer has any idea where he is, given he was closed up in a box throughout the voyage. Jaunet feels close enough to the humans in order to calm himself down and stay, but not close enough to accept the changes, to adapt and establish a new territory. As such, he merely saunters from "one nook and cranny to the next"—an inactivity that today would be read as the result of extreme stress—and he subsequently breaks all friendly contact with humans; he no longer rubs up against their legs, no longer purrs in close proximity to them, *in spite of "being well-treated" by a good and attentive family*. He adopts a wild gaze and a somber demeanor," according to Fabre, who reads and translates Jaunet's reaction to humans as a depression, which we now know to indeed be possible for cats to undergo.

Through his gestures, Jaunet expresses, adjusts, and reconstructs his representations of these humans, incorporated into his world. But they are neither the main focus nor the centerpiece; rather, they are integrated into his territory but not to the point of changing it or whereby he would himself change or overcome a change. He responds to their configuration, since they are no longer the same within this new environment, no longer enveloped by the same scents, sounds, and landmarks, and he as well changes his configuration in changing. Jaunet ends up dying several weeks later, unable to overcome his "misery" and "sorrow," unable to reestablish some kind of inner balance (homeostasis). One morning he is found lying in the ashes in the chimney, where he no doubt would seek out warmth during the night since his early adoption like many a house cat during that era. His failure in survival is a result not so much of the territorial nature of domesticated cats but of his territorial culture, of the way he had shaped his way of being, through his epigenetics, in contact with the environment he found and then inhabited in Avignon, an environment that he also, in his own way, slowly but surely shaped and organized.[11]

A similar milieu favors similar cultures within space and time, in particular between generations of the same lineage from the same place, as indicated by one of the younger of the Jaunets. The cat finds himself captured and gifted to several of Fabre's friends a few days before their move

in 1870, then enclosed ("in a shut basket") just as quickly, pupils dilating [*écarpupiller*], surely quiver-sniffing, perhaps licking to taste (wicker), no doubt scratching and meowing, already stressing out, having never been enclosed before. — . . . paNIC when he senses the ground move and stir: BRACING HImself, CLAws planted BODY LOWERED HEAD TURned IN WHiSkers EArs FOlded back FuR rAISED On End EYeS DiLATEd;
unPLEaSanT surPRIsE . . . —

No doubt he probably meowed and stirred incessantly (heading quickly to the nearest neighborhood) while seeking to quiver-sniff a nearby scent that was familiar (Fabre), ear-turning toward familiar sounds (if he responded in order to calm him), which perhaps reassures him a bit, allowing him to discern (between the wickets) smells and noises encountered and passed throughout the penumbra (transport *at nightfall in order to incite* the animal to transition into his new home). Upon arrival, *the cat is left isolated in a bedroom, as is suggested, in order to force him to run around, orient himself, and become accustomed* to his new home.

> You have perhaps noticed that, in order to help us to decenter ourselves—you as well as me—and allow us to get closer to the animals, a literary process can be used to invent words and expressions, utilizing and emphasizing certain behavioral characteristics.[12] Hence the terms "ear-turn," "quiver-sniff," "palpitate," "pupillate," etc., rather than "listen," "smell," "touch," and "look," all of which are very abstract. Such invented expressions allow us to emphasize the animal side: how he acts and lives. These poetic attempts are concerned with the body and gestures. This doesn't mean that cats can be reduced to this, but this is what humans are best apt at grasping. Other aspects of cat communication—their scents, pheromones, vibrations, micromovements—are all less perceptible to humans, at least consciously. Conversely, words still need to be created for the sounds cats emit. However, the eyewitness accounts do a poor job of describing them, since humans can hear them but poorly distinguish between them (at least for the time being?), either by negligence or difficulty, contenting themselves to use the ideophone "meow," which is expressed differently in other languages.

— . . . feels himself climb down, settle down, and stop . . . de-pupillate the light . . . launch his paws . . . perceive the unknown surrounding him . . . except for one [Fabre] disappearing . . . grasping that he cannot follow him as he leaves . . . — "As soon as he feels himself to be a prisoner in an unknown room, he sets off leaping around the furniture, onto the windowsills, among the décor of the chimney, threatening to make a mess of everything." No doubt. — . . . rounded back, fur on its end, head ears raised yowling spitting disoriented inside the unknown space scared by being closed in and heading toward the areas where light and sound permeates the room from the outside . . . —

This vagabond from a specific territory behaves like his elder, but he is not able to calm himself down by seeking out the Fabre family. He leaps as soon as he senses odors and air coming from the outside and jumps and passes by the part of the room where they are entering: *a first floor window is quickly opened by the frightened owner, who is not accustomed to these kinds of cats perhaps and quickly becomes frightened himself as many others did during that time*; —scAmPEr aWaY . . . follow the trail of scent, the exhalations of the human, spread-pupillate sLINk ANd wEAve avoid stationary obstacle as one moves one's legs and the dogs with their rotten stench . . . MOvE ones paws . . . suddenly liquid obstacle in front of scents to be captured farther along . . . traverse . . . GaLloP full speed ahead recognized scents recognize places reassured movements . . . emanations and noises from the locale, of the humans living there . . . l_e_a_P . . . "We saw a large body of shimmering water leap onto the windowsill," wrote Fabre, who deduced that the cat had traversed the Sorgue by swimming. He perhaps hadn't seen or thought of the bridges, the various scents and landmarks used to navigate having been attenuated or lost, or perhaps he wanted to take the shortest route possible, pressed to relocate his territory, driven by fear, led by his own sense of direction—a faculty that has been observed in cats for a distance of a dozen kilometers but which has yet to be explained, and which in his case was not based off the sun since he embarked on his journey at night.

"This formless bundle came rubbing up against our legs joyfully purring," reaffirming his markings and presence, no doubt also doing the same thing with the salient elements of decor around the home, in each specific room, slowly reassuring himself. Since he demonstrates his "loyalty to the residence," *Fabre decides to take him along with them, but he makes an in-*

terpretive mistake and a bit of semantic confusion arises: Fabre thinks the cat has a loyalty to the family residing in the residence, whereas it's perhaps better to think of the cat having an attachment to the residence itself of the family. The latter being simply a salient element within a much larger territory, which is the primary element; it is not a primary relation.

> If you confess to your veterinarian that you believe your cat rubbing against you is a sign of affection, there's a pretty good chance he will raise his eyes and arms up toward the heavens and exclaim: Too much anthropomorphism! Too well informed by their biological instruction, based on genetics and physiology, most veterinarians prefer a biological . . . explanation. As it has been observed that adult male cats rub more often than female cats or kittens, it is understood that they only perform this marking for spatial reasons combined with a relation to domination. However, the human is not an object, but a being replete with emissions and vibrations which a cat can feel and differentiate very well. This same human is also a being of interactions, in this case through food and petting. It is therefore more judicious to think that the practice of rubbing against humans by cats eventually gained a social, interspecies and emotional function following the same trajectory as purring and meowing. This is a cultural function, since not all cats live or no longer live with the same intensity within space and time. That is not to say the emotional dimension always wins out over that of territorial marking. It certainly seems to be real, but secondary as far as Fabre's cats are concerned. Fabre's cats are merely one very unique instance of how cats can adapt to the living conditions—residing as they do within a very specific and localized area, place, and environment.

The younger Jaunet would have no doubt lived along with his ancestor in Orange if he hadn't been poisoned not too long after the family moved (*a common way for cats to die during that time, as humans sought to rid themselves of stray cats and dogs by any means necessary*).[13] Or, perhaps he would have lived like the third male cat of the Yellow Cat family, of the following generation that *the Fabre family decided to bring along with them in their move from Orange to Sérignan in May 1879.*

But he as well found himself snared in a trap, raised up and tossed into the darkness of a basket, feeling his world suddenly move and sway beneath his legs. He must have been scared and frightened as well, ears head tail lowered claws at the ready, twitching along with the mobile ground, on alert for noises below and the hammering beating of horse hoofs ahead of him. No doubt, leaping when the basket was really shaken, regularly meowing, clawing at the lid every now and then in hopes of escaping. But through the wicker basket—his ears pricked, nose wiggling, eyes pupillating—he perceives sounds scents coming from known humans (the Fabres) as well as from other male and female cats without, however, detecting other male cats (that had been offered to them over several generations or just prior to this voyage). This probably reassured the cat and allows him to control his stress since Fabre signals to him that nothing strange is happening (they'll be traveling seven kilometers over one or two hours).

As soon as he feels solid ground and he de-pupillates to the light, he finds himself captured again and then released (*in the attic where Fabre intends to keep him for eight days in order to get him used to his new environs*). Pushing his paws here and there, pupillating the light, quiver-sniffing the scents, "paw padding the ground," ear-turning toward noise, fur-vibrating his whiskers: an enclosed space, without any known markings from him or his peers or those humans close to him. In contrast to the prior male cat, he does not panic, possessing a different, calmer demeanor, as well as being reassured by the regular presence of a known human (the Fabres have decided to carefully check on him) and occasionally encountering the presence of a female cat (*brought to their new abode later*). He can eat more than before, "more plates to lick." He feels like he's getting more attention and more pets from the humans because *they want him* to recognize the advantages he has here and forget the past.

Once he lets go of his defensive postures, LOWERED HEAD, EARS AND WHISKERS FOLDED BACK, stiff curved BACK, tail WhIPPIng, he finally makes himself "soft to the touch," more than likely with an arched back, tail and ears pricked h-e-a-d-t-e-n-s-e, whiskers narr'w'ed, and when he runs after being called, he rubs up against Fabre's legs and purrs. Fabre decides to release him and hopes that this attachment will override any attempt at fleeing. The cat descends the stairs, more than likely behind a human, realizes that there are even more new forms and new scents to discover, such as the kitchen, where the presence of other cats perhaps reassures him a bit, and such as the garden, where he runs around before returning inside all by himself, *leading Fabre to believe that the adaptation has taken,* whereas the cat had merely made use of all his senses to grasp something of the unknown expanse, the loss of his world and therefore of self.

Let out the next day, he disappears just as quickly— . . . fleeing . . . found again . . . following his path . . . pushing his paws along the earth . . . getting them wet in the cold water . . . —according to Fabre, who finds him "covered and crusted in red dirt" and yet dry and therefore deduces that the cat must have not followed the road and taken the bridge, but veered right and swam (in the Aygues)—shaking himself dry "in the red dust of the fields" . . . quiver-sniffing then pupillating the various plants, stones, the entrance, his own markings . . . —he is found "meowing at the foot of the locked house," indicating that he doesn't see the Fabres as a distinct entity but as an integrated, assimilated element, belonging to his territory, that they cannot live elsewhere. He therefore lets himself be taken by *them, having returned,* but immediately finds himself confined once more within darkness (inside the basket) then released and left alone in an enclosure (he is left in the attic for two weeks). He behaves calmly and doesn't show signs of aggression, nor urinary markings, nor compulsive agitations, nor any signs of stress as would be the case with contemporary cats, but it is possible that he has merely inhibited himself, or overcome such stress, and he voluntarily interreacts, becoming more singular, distinguishing himself even more from the preceding males of his lineage, one being constantly stressed, the other panicking.[14]

> Such an episode allows you to see the importance of individuality, defined as the collection of singular traits by which an individual differentiates themselves from others. This is something that has only recently been discovered by ethologists and veterinarians. And yet, such a process structures each being, since these expressions can be located all the way down to the biological strata, such as in plasma, which has been used to legitimize recent techniques of profiling in order to trace and prevent certain feline illnesses. Individuality can be included along with genetics, epigenetics, or the environment, whether in the case of humans or cats, in order to provoke variations between individuals. I put forth the notion that this individuality is both biological and cultural, since it is also shaped during the moment of infantile initiation and throughout the course of one's life, during the end-

less adjustments to one's environment, and it produces what psychologists and ethologists refer to as personality (a set of particular behaviors). Individuality explains why cats acquire different personalities within a common environment, such as here with the Fabres, or even converging personalities within different environments. But there are limits: confronted with other outside factors, dissimilar individualities can even generate quite similar personalities under the pressures of a very intense environment.

Ethologists are also interested in temperaments, which they differentiate from character traits, two notions that are used to denote given psychological and behavioral dispositions already expressed at birth. Whereas character traits, which are indeed individualized, had only been attributed to humans, temperaments, thought of as statistic frequencies at the group level, seemed to suffice for animals considered to not be individualized. In my opinion, the recent focus on the individuality of cats seems to invalidate such a distinction and should lead us to speak about character traits as individual expressions, even if they can be regrouped into more or less frequent types: as temperaments.

As for personality, ethologists have begun to grasp it as a much more complex process. Each individual forges a personality during their process of socialization with the environment, adjusting and modifying his or her character traits and temperament. The personality is not a given but a construction, a result. As with other ethologies inspired by human psychology, feline ethology is now attempting to measure a cat's personality. Calculated through measuring the expression of five behavioral dimensions, such measurements provide a photograph of a group, period, or place and situate each individual within said group, with his or her own individual particularity.

However, a good number of ethologists believe in a process of rapid socialization at the infantile stage and a subsequent fixed personality not too long after. Insisting on the constitution of self through action, the human sciences have provoked us to envision the construction of personality throughout the course of one's entire life, due to the endless activity of the individual within the environment and if there happen to be adjustments to this environment, then we should also see adaptations and modifications of the personality. The personality would therefore not be something that is forged only during a specific period of time but rather a provisional state, of a personality undergoing endless change, constantly being adjusted, modified, and forged within and under the influence of the environment. Rather than the programmed, initiated being of the ethologists, or the being that merely constructs itself according to anthropologists, such a biological-cultural conception of personality would give way to an equipped being, initiating itself, deploying itself, adjusting itself, and perhaps transmitting a part of what it has learned through epigenetics.[15] Hence the insistence on the singularity of individuals and their trajectories, without forgetting other behavioral factors, such as gender and age.

This should explain why I have focused my attention on the historical eth(n)ology of individuals when writing this book. For it is individuals who acquire, bear, transform, and transmit environmental characteristics. It's thanks to individuals that we are capable of grasping most easily the construction of behaviors and are able to analyze them. It's within the individual that the links between cultural and biological dimensions are embodied, the past and the present, the living and their environment. And furthermore, it's precisely concerning individuals that we find the most forthcoming and well-articulated historical documents: human witnesses, often themselves cat owners, but not always; above all, they are good observers who pen good descriptions in memoirs, journals, fictional tales, and biographies.[16]

However, unlike my work presented in *Animal Biographies*, in this present work I focus less on individual animal lives or a cross-section of an animal's life. Instead, I seek to concentrate on portraits and situational points of view and the action of animals within their environments. By taking this viewpoint I can demonstrate how these beings construct themselves within a given environment. In this way I can provide a historical view that shows to what extent these beings live in nature in a much more fluid manner than humans tend to think, since such a claim has been endlessly repeated in the West. Rather, these beings live within their cultures and are endlessly adapting and thereby rendered dynamic.

Like the cats that were moved in 1870, the cats from 1879, possibly the same cats or their descendants, adapt better: "They explore the rooms one by one; by way of their pink noses, they recognize their furniture: the chairs, tables, and armchairs are the same but the places are not. There are small meows of surprise along with quizzical gazes." They are abundantly fed and petted, *the family wanting to reinforce and retain their connection and the cats acclimate quickly*. The female cats, which the Fabres let live together in a collective with their kittens, have no doubt developed a strong sociability between them, perhaps even a cooperation inciting them to stay together, to not leave the group, which has now become as important as following the footsteps of their kittens. They must have also integrated the Fabre family into their group, *an assimilation reinforced by petting*. And thus nursing mothers must think it's practical, economical, and convenient to be fed without having to expend any effort, and the subsequent surplus rations they receive from the Fabres only serve as an additional reassurance and reinforcement of this impression.

In complete contrast to the mother cats, the male cat roams for several hours in the villa and the garden and surely confirms that he is not in his territory any longer and disappears, returning to his home. Left there *by the Fabres who have let him go*, and now indeed finding himself within his territory but in a different environment since he is no longer fed, he modifies his way of being, his personality, and becomes a hunter of the farmland and barns in the countryside surrounding Orange, having been "seen one day behind a hay bale with a rabbit between his teeth," and therefore doesn't hesitate to attack large prey and take risks while also widening his living space as is common for contemporary cats, even though he had hardly hunted at all before, according to the Fabres. He therefore adapts in his own way, without more than likely adopting the codes of other farm cats, who still today hunt rodents, insectivores, and birds, but not the animals that are raised by the farmers in order to be tolerated and survive. *Furthermore, without any other news about the male cat, the naturalist presumes he has been put down as a dangerous nuisance, as was often cited in literature of that era.*[17]

These rescued cats don't seem to be that numerous during this time, since their humans are still a minority. During the time of the Jaunets, the typical house cat is more than likely a farm cat or a cat of the village, which elaborates, develops, and maintains another type of territorial culture, in the image of . . .

Toto: Held at Bay

(Bex, 1873)

> . . . described by Athénaïs Michelet (1826–1899) who, more partial to city cats accustomed to the living room, provides us with a rare documentation, since anecdotes about cats are hardly widespread in the literature of that time. The second wife of the historian Jules Michelet, collaborator on the books *L'Oiseau* (1856) and *L'Insecte* (1857) and author of *Nature* (1872), Athénaïs pays careful attention to animals and in her off hours she writes a work on cats with whom she very quickly entered into close contact, as was already noted in her earlier work *Memoires d'un enfant* (1866). Constantly interrupted, the editing work on her cat book was still not completed upon her death in 1899, but the manuscript was published as it stood in 1906 under the title *Mes Chats*. The work relates the ties woven with the cats she took care of or had encountered, including complete or partial histories of everyday life, some consisting of daily accounts. Athénaïs distinguished herself as a field ethologist before her time, like Marie Dormoy will be for her cat, or later on, someone like Colette Audry for her dog, already possessing the necessary qualities presented and emphasized still today by ethologists, such as good listening skills and concrete empathic observation. The skills requisite for providing a significant amount of information, in spite of the romantic gushing specific to her era. In particular, she writes magnificently regarding village cats of the Vaud in Switzerland, which she observes while on a vacation with her husband in the summer of 1873, and which, hardly every closely studied, live within a fluid context of status, reinforced by their territorial culture.

If they don't end up getting themselves drowned, unwanted cats, born in these village homes, are eventually led into the forest and abandoned, high up on the Swiss alpine slopes in the areas left for the vineyards, leaving them to fend for themselves among the birds and rodents, *which presupposes that the human decision-making involved in such cases takes place a couple of weeks after their birth, after the kittens have emerged from their hiding places where the mothers had sheltered them from humans, and as the latter discovered them and weren't able to find them a family to live with in the surrounding area.* Alongside those who will die quickly, having not learned how to hunt or do so poorly, and those who will definitively remain strays, *but which Athénaïs does not mention, there are also* those who know how to become (re)adopted *by old folk or the children who encounter them in the woods while out gathering dead tree limbs for firewood.* They approach the humans with their tails held high—a sign of confidence—arching their backs in order to solicit a friction, touching human legs to then rub them in order to mix scents, *according to the writer who interviewed their owners*. Returning from the wooded mountainous regions, settled back into houses or barns, they retain their desire to revisit the woods where they had once marked their territory, often following their owners back there when they would themselves return to seek out more firewood.[1] As other cats also find their way to these villages, the felines of these areas live a precarious life and a fluctuation in their living conditions facilitated, *justified*, and maintained by way of their lack of contact with humans, through a distance that *the cats themselves impose*. The distance is initiated and inculcated at a young age, even in those cats who are kept by humans, as is noted in the following commentary about a grey and white cat, common to this region, born on a neighboring farm.

There, as in the surrounding areas, the entire litter of kittens is kept (not merely out of indifference, as Athénaïs thought was the case, who would like to see a reduction in the number of cats and for those that are kept to be well taken care of) but in order to let life and death settle the score between the weak and the strong, as with humans. As for the exchange value of cats and dogs—it's nonexistent and little attention is paid to such an idea, given that both animals can be replaced at will. The grey and white cat in question had just been given to the manager of the Bazaar at Bex in order to help keep the mice at a good distance from his boutique, above which Athénaïs stayed in her hotel. And it's from her hotel room above the bazaar that she was able to observe the daily movements of this feline, establishing its age as three weeks upon their eventual encounter.

> But I will stop you right here! Re-read the last words in the sentence above! The autonomy Athénaïs is describing doesn't fit the description of common contemporary cats of a similar age. On average, kittens don't even open their eyes until the age of two weeks, only demonstrating an efficient thermoregulation, a more or less fully developed olfactory capacity as well as an ability to focus their vision and hearing. Their capacity for orientation begins around the age of at least three or even four weeks old, and the ability to potty independently of the mother begins at five weeks, etc. I could posit that Athénaïs is mistaken in her estimation of the cat's age and gives the cat a younger age. Such a speculation would lead to casting a cloud of doubt over her entire documenta-

tion. This would certainly be tempting for those ethologists who believe in the intemporal character of contemporary cat behavior (since they consider demeanor as specifically biological) and who remain suspicious of popular anecdotes, especially if the narrative arises from a past suspected of being ignorant. However, my experience as a historian has led me to search elsewhere in order to grasp an even portrait of cat behavior described by humans of this era. The people from this time period clearly know how to observe nature and Athénaïs has been around kittens since she was a child. Even if I recognize that Athénaïs's descriptions are not chronologically correct, that is, her writings on the cat are merely a collection of interwoven comments from the cat's early period as a kitten, it's still not enough to explain the age disparity and behavior. I have two hypotheses: one individual and one specific. Separated from her mother and sibling very early on, the kitten must have had to quickly learn to fend for itself to survive, which forced it to develop more quickly. Even today, such a situation would generate significant differences in a cat's behavior and development. Kittens that have been cut off from others can afford to be more precocious than others. And more generally speaking, wouldn't cats in this earlier environment not become a bit more mature at an earlier age given the difficulties they face compared to their contemporary peers? Humans have experienced such situations. And facts would indicate that felines also have a biological plasticity that would allow for and even suggests this early maturing process. Hence a quicker time to procreation for wild cats living in a harsher environment. Or the opposite: the recent lengthening of the average lifespan of domestic cats living in an easier environment, which could lead to an expansion of their period of development and thereby yield less precocious kittens than their ancestors, or at the least for some among them. For the moment, we can only speculate, but let us, you and I, have confidence (obviously relative) in Athénaïs and examine her observations more closely.

This cat just endured the violent erasure of her mother as well as the violent end to her breastfeeding, which should have continued. She is certainly disturbed and is dealing with strong emotions and physiological tensions, to the extent that she also feels herself diminished in confronting the world, unable to partake in it through observation/imitation like her peers, but rather forced to do so through a difficult series of trial and error that are all the more aleatory, disconcerting, and dangerous. In the beginning, the cat is able to compensate for her loss of a mother thanks *to the children in the bazaar who constantly solicit her attention. Such a situation leads to an* acceleration in her development, as can be observed in contemporary cats who have also been manipulated prematurely *by humans. However, such accelerated development by way of human contact only leads to cats developing behavioral aspects that align with the desires of humans themselves, in this case, for entertainment such as playing.* For the grey and white cat responds very well to the initial "constant calls" and she "is overexcited in the games of all kinds, making the children laugh, surprising them by her gestures," thereby reducing the fear of these beings, learning, inventing even her response to the solicitations she experiences as stimulations. Perhaps she even begins to respond to the name Toto, *which was more than likely given to her by the children during this time, who were hardly thinking about the cat's gender but were seduced by her attitudes, which were viewed as farcical.*

> Some among you will exclaim that the term "attitudes" is reserved for humans, that instead we should employ the term "behaviors" for animals. And it's certainly true that since the beginning of the twentieth century, ethologists have defined the actions of animals in response to their environments as such. This is the case for two reasons. The first is a certainty and desire that such responses are of a biological order, and therefore that such behaviors are identical for every cat of the species, even identical for the rest of the animal world. This is a conviction that "attitudes" or "conduct," which implies choice, are reserved for humans alone. Rising up against such reductionism and anthropocentrism, certainly more prevalent in past eras than today, zoo-sociologists propose to use the terms "attitudes" or "conducts" to signify when actions are constructed (and not imposed) at the level of the social or cultural, and not the biological.[2] I share their opinion but not without disagreeing on certain specifics, in particular regarding what is constituted as biological or sociocultural. That is, the most habitual of behaviors have a social dimension. And the most singular behaviors have a biological dimension. So, I deliberately use the three terms: "behavior" for habitual actions, "attitudes" for those actions adopted according to cir-

cumstances, and "conduct" as a response to a precise conjuncture as choice or singular behaviors.

However, the cat's possibility of attachment and interaction is quickly countered by *human customs* and the cat's adaptation to them. *Taking on the role of mother to this orphaned family of cats, the oldest sister gives the grey and white cat a "dish of cold milk with bread every morning." This meager morning ration at a fixed hour is commonplace for rural families of the period and willingly encouraged by many in order to dissuade the cat from begging and to get them to turn their attention quickly to their work at hand: hunting mice, without worrying about whether it is too early or not (today, kittens generally don't kill mice until the age of five weeks old)*. Rather, they would on the contrary like to hurry them out to hunt, to set them to work, prepared to lose the ones who are too weak, considered useless. As such, the oldest daughter receives poorly the spotty advice from Athénaïs, who has a more cautious, bourgeois conception of life: that of adults, children, and animals. The first won't give in to the second's request to give the cat meat, which is already quite common in bourgeois households but which remains fairly irregular in more modest households, in which it is unthinkable to give what little meat one has to animals. For such a gesture would also deviate from the larger plan of getting the cat to quickly learn to fend for itself, *as is often demanded of children as well*.

Besides experiencing the incomprehensible disappearance of maternal breasts and their lukewarm milk, she will from now on only perceive the cold milk from another species, for which she will at first feel a profound disgust. In a similar manner as the finicky cats of today, this young cat remains hungry confronted with this less digestible, less filling beverage, which is above all less abundant, in spite of her frequent demands for it. For she quickly makes a connection between the sounds, scents, and gestures of the morning arrival of the older sister, the sound and movement, the subsequent sound and movement of the opening boutique door and the placing of the full dish on the ground: — . . . starving and impatient, does not wait to precipitate moving her paws toward the human . . . circling her meowing, raising her paws clinging [dress] lowers her paws meows turns meows some more and raises her paws up again— . . . thereby signaling her needs, desires adapting to her *interlocutor*, hoping to be understood, *which is certainly the case since cats and humans are in close enough proximity and community for that*, but without having an effect on the quality or frequency of receiving the milk as she negatively experiences the disjuncture between receiving multiple feedings of her mother's milk throughout the day—which kittens adore—instead of this lone distribution.

And, little by little, she will also repress an "instinctive aversion" for this unknown food and more than likely will begin to signal to the oldest daughter not to change it but increase the amount given to her. The cat will return to the dish earlier and earlier in the day, discern that it's empty, seek out and look at *this human*, approaching her with "a subtle meow full of sorrow, half complaint, half prayer," *Athénaïs reading and translating her expressions with human words*. All in vain. The cat will also arrive faster and faster, quickly swallowing up the few morsels *the writer offers her. Officially visiting the shop in order to buy something, Athénaïs makes use of her time to give something to the cat to eat instead, countering the local customs*. Is the cat already hunting mice? *Athénaïs doesn't say, estimating that it wouldn't be enough food anyway, also in contrast to local opinions*. Perhaps she has difficulty hunting, separated too young to have seen her mother bring back living prey and release it from her mouth in order to show the kittens how to grasp it in theirs.

The cat is confronted daily by the oldest daughter, who doesn't respond to her bodily addresses nor even look at her—and more than likely doesn't speak to her. In no way providing what the cat is asking for, the daughter forgoes the possibility of serving as substitute mother for the cat—*if she had responded* to her. And as a result, the cat must have slowly felt a distancing that she reinforced in seeing herself isolated on a daily basis, *left to her own devices at night in the boutique to mind the nocturnal rodents*. In this light, the cat no longer received any socialization skills from her mother or her surrounding siblings and hardly any socialization from humans. Starving, she becomes all the more tired and thinner as the days go by and shuts herself off: she sleeps more to retain her energy, doesn't move around as much to save her strength, slowly circulating. Even though she gets excited when she *sees the children offering up their caresses*, she quickly tires and stops. Here, we find Athénaïs providing details of how the cat reacts, which would correspond exactly to how contemporary cats would react if they were separated prematurely from the mothers and were then asked to play. She therefore provokes the *withdrawal of the children: "the latter having first initiated calls of friendship, but seeing that they had become useless, she detached herself from them and no longer responded to anything."*

Reacting and *adjusting the environment to her own needs*, the cat adapts, both willingly and through coercion, slowly but surely creating her world, within which *humans interact less and less, which was what was desired from the very beginning*, fixing her (fairly restrained) territory of the cold and humid boutique to the street, where she sunbathes to compensate for her lower nocturnal temperature, physically living with *these humans*, sharing the space, scents, noises, and movements, but at a distance for emotions and gestures. Slow but surely, she shapes her way of being, her individual culture—suggested, imposed, and transmitted—*by these humans who have a goal and relation in mind*. She could die, *as Athénaïs fears*, and which is often the case with a good many contemporary rural cats, if they don't end up adjusting themselves and turning toward eating mice. Or, she survives thanks to the latter, settling into this new life, maintaining a cold coexistence with these *humans who precisely want such a relation*, reinforcing this direction of her being and world, becoming a cat of the Swiss village of *Bex, of a house and street*, integrating into a feline population educated in a similar manner, composed of individuals strong enough and well adapted to being able to survive: *according to Athénaïs*, they consist of relatively healthy males and females constantly tired from pregnancy, kittens who are quickly separated and given to others, who must then adapt and adjust, forging themselves, dipping deep into the epigenetics and confronting this demanding environment. Gathered up by Athénaïs for a while, Toto survives and then is placed with a family. *We don't know anything else, as the Michelet family left the region.*

This type of cat, its way of being, its individual culture, can still be found today throughout the countryside, most notably on farms where their utility is still solicited, taken under the care of humans who, as a result, provide them with an analogous environment.[3] During this earlier time, cats from the countryside also lived in houses: in the locales of modest city-dwellers and the bourgeois, but their importance and singularity has been poorly documented since they have been forgotten by the literature of that period save for the rare occurrences in the work of Léautaud. It was impossible to glimpse the individual and daily culture of Fabre's cats, as he almost never spoke about it and already chose to live with a fairly specific type of cat companion: cats which were simultaneously adopted, but left to stray, and whose relation was intermittent. Such a daily description of individual culture of cats appears more clearly with . . .

Moumoutte Blanche: Muted Presence

(Rochefort, 1880–1889)

> . . . who embodies and illustrates another variation of individual culture in the family house of the French writer, Pierre Loti (1850–1923). Bearer of a luxurious white coat of fur with black patches on the top of the head, between the shoulders and tail, the cat is an Angora mix. The pure breed Angora is completely white and had been introduced into Europe in the seventeenth century, designated as a cat of the aristocracy and viewed in contrast to the vulgar black cat, and eventually became a minority under the influence of uncontrolled crossbreeding with the rest of the feline population. This was largely due to its social diffusion and more broadly to the sociological extension of the nineteenth-century aristocratic model of the indoor cat [*chat du salon*], principally in northwestern Europe and North America. And by the end of the nineteenth century, entire treatises on the education and care of cats, children's literature, even a newspaper, are written for this specific public. This new literature wipes away all the previous criticisms and warnings concerning cats that were still largely present up until the middle of the century and amplifies another image from much earlier that glorifies the cat's elegance, beauty, and intelligence, even raising it to rival the place of dogs, still promoted as man's best friend, affirming their attachment to people, but conflating, like Fabre, this attachment that is in fact an attachment to their residence. Promoting the cat of the salon, this literature promotes breeds that are already accustomed to such a life (the Angora, judged to be beautiful, the Chartreux, and the Iberian Lynx), all quite different from the common black, grey, and tabby cat varieties.

However, the parallel proliferation of street cats during this same time also led to a profusion of cat portraits in the last third of the nineteenth century: freedom and independence are attributed to the most elegant cats, beauty and cleanliness to the more plebian cats. Hence the emergence of an ideal, uniform, and eternal cat, accumulating all the positive attributes from each breed. Common crossbreeding, which one can't or does not want to prevent, becomes justified and reinforced, all the way to seeing certain mix-breed Angoras among ship rat hunters such as Trim, whose tale we will recount later on, or among the stray city cats, like those discussed by Léautaud. This tendency is counter to the standardization and officialization of breeds, first and foremost those considered elegant, through the momentum of feline expositions (United Kingdom:

1871; the United States: 1895; France: 1913), in an attempt to preserve them but with limited effects.[1]

Julien Viaud, more commonly known by the pen name Pierre Loti (which he used in order to circumvent the military authorization necessary to publish), dedicates an entire chapter in *The Book of Pity and of Death* (published in 1891) to his cat, Moumoutte Blanche. This volume is the fruit of an autobiographical work that marked a significant portion of his other works, which he will continue to develop following his work *The Story of a Child* (1890). Loti's novel appears quite faithful to reality, as is verified through his descriptions of the house with blueprints from the latter work. However, he evokes the period 1886–1889, and makes no mention of the renters who were living there up until 1884. Nor does he speak of his wife, whom he married at the end of 1886. And even if Loti places himself at the center of the work, the reality was that he was largely absent from the house where the cat lived during his years as an officer in the French marines. For the cat lives mostly with Loti's mother and aunt, whom the writer refers to as Claire, but whom we will refer to by her real name, Clarisse. According to Pierre, Moumoutte Blanche spends even more time living with the butler, Sylvestre, starting in 1896, but the latter's name hardly figures at all in the pages of the novel. Nevertheless, Loti displays his desire to describe the animal's daily life, and while he is less precise in his descriptions than Athénaïs Michelet—more focused on himself and his own reflections—his writings are still of great interest for us given that such firsthand documentations of cats are very rare from the era. One can verify and enhance such description by referring back to other confessions made by the writer: for instance, one can refer back to his journal, which served as the basis for a number of Loti's novels and autobiographies, such as his *Roman d'un enfant en prime jeunesse* (1919), a novel that, in spite of the chronological gap between the periods depicted, provides us with vibrant descriptions of his home, the atmosphere, and the ways of being of the humans residing there. In other words, Loti's writings provide a sound description of the environment in which and through which the animal structured itself. This approach is reinforced by the conservation of this house, which still exists today in the same exact layout, allure, and décor from the era, along with available photographs of the inside taken by Clarisse and the writer. Moreover, one of the photographs even shows Loti holding a dark-colored cat in his arms with a white one resting at his feet, which must certainly be this Moumoutte Blanche. Loti also has a self-portrait painted in 1891 by the artist Henri Rousseau, which includes a tabby cat. For it was commonplace for the bourgeoisie of the era to adopt the aristocratic practice of having portraits painted with their cat or paint-

> ings of their cats all by themselves acting out scenes from their daily lives. As such, we see the scholar Ernest Renan display a painting of his mother-in-law's cat. This genre of painting will remain very popular throughout the twentieth century and will be reinforced by photography for a bourgeois society that appreciates realist art.[2]

The cat seems to have arrived at a very young age into the house of Pierre Loti in 1880, while Loti himself was still out navigating the Adriatic Sea and stationed in Montenegro. No doubt it was given to him by friends from Rochefort, some of whom had previously had Angoras. Officially the cat was acquired in order to hunt mice in the salon; then, since these Angoras were considered to be mediocre hunters, it probably became a good companion for his mother and aunt in order to make its presence felt, but without either of the women admitting to it, since later on they will partake in the burial of the cat in 1889 out in the garden, without making a big deal of it, chatting among themselves on a bench since one shouldn't confide too much in others in such an austere protestant milieu, less still at an occasion dedicated to an animal: "To meet expressly for this burial of a cat would have appeared, even to ourselves, rather ridiculous."

The cat roams about the house and discovers three distinct levels, each consisting of spaces in rows, starting at street level and following a long narrow hallway, a disposition that invites the cat to endlessly climb and *bound down the steps of the two spiral staircases, to* traverse and really get to know these spaces, seamlessly moving from one to the other in order to move about the house. She doesn't encounter humans that often due to the family's financial problems, which led to the firing of several maids prior to her arrival, along with the death of grandmothers and great-aunts, aunts, a brother, and Loti's father. There are hardly any objects—the remaining surviving relics that hadn't been resold to dealers attract the cat to rub against them, to stretch out (the carpet), or climb up (tables, chairs, chests of drawers, a writing desk) in order to look down from above, to better protect herself, reassure herself, and be on the lookout—the desire of a predator—and avoid the cold floors. Roaming around allows her to get a feeling for the textures through her paw pads, head, and rubbing of the hindquarters, helping to learn and differentiate the various spaces and objects in a tactile manner: the smooth waxed surface of the wooden floors with furniture everywhere, since "immaculate *cleanliness and extreme order were the sole luxuries,*" the coarse woolen surface of the rugs in the living areas, the soft velvet of the chairs and upholstered armchairs, the slippery cold surface of the terra cotta tiles in the kitchen or the marble of the chimney, the dry cotton texture of the tablecloths.[3] She more than likely also distinguished the rooms through their luminosity, produced by their colors and the light.

> You perceive cats, your cats, with your human capacities. Don't believe for one second that such capacities are normal or better than a cat's perceptive capabilities. Never forget that cats, *your cats*, perceive you through their own modalities. Every animal species, including our own, possesses particular sensorial dispositives, which provide a sensorial framework demarcating what they see, hear, smell, and feel, or don't, and which allows them to perceive what they are, in their own way, capable of. It's precisely within such a sensorial framework, alongside their cognitive framework, that each individual apprehends their environment, through evaluating, acting, and reacting. Each individual sees the world in her own way and evolves in her world. However, the account of the latter is not simple, since the modalities of perception are poorly understood, even less than the postures, gestures, and actions expressed to no one in particular and therefore observable, whereas these perceptions are interior and are tied to cognition. Ignorance concerning cat behavior remains strong, in particular regarding their vision and perception of colors. The question of colors is a good example of how Westerners (but they aren't the only ones) have thought about others. For felines, as for a number of other animals, researchers have gone from holding on to the certainty that they only see in black and white—so as to leave the presumed superiority of seeing in color to humans—to eventually conceding that other animals do see in color, but without coming to any clear agreement on what the colors are, due to the different ways felines can be observed and examined. And ethologists often reason by way of privation—cats do not possess the ability to see this or that color—through recourse to human perception taken as the absolute norm, when in reality such perception is merely particular. In contrast to this, we should think of cat perception in terms of originality and difference and ask whether or not cats perhaps see other colors humans do not see and perhaps other shades as well. So, be wary of restrictions that often serve to construct a hierarchy, especially considering Westerners have done the exact same thing

with other human populations (but these initial biases often culminated in recognizing identicality). In 1949 a newspaper from Brussels was still reporting that "Black people could only detect three colors"![4] That being said, in the absence of any certainty on the color spectrum perceived by cats, I will merely reflect on the criteria of luminosity, which is much easier to think about.

— . . . drab areas on one side [the street on the east side] . . . more luminous on the other [the southern, courtyard side] . . . with variable light depending on specific moments [the seasons and positions of the sun] . . . darker up on the upper floors [bedrooms with blue and green tapestries] . . . brighter below [living areas decorated with lighter colors] . . . with some reversals . . . the lower levels are darker in shade [a salon is painted in red in 1885, the dining room becomes a Japanese pagoda with filtered light in 1886] another room upstairs is more illuminated [the master bedroom is painted white after Loti gets married in 1886] . . . — The cat must modify her luminous cartography but without much trouble or consequence, since cats adapt their sight to variations in light intensity. Moumoutte evolves in an attenuated luminescence (*often desired and accentuated during the 1880s). Loti overloads the walls with exotic souvenirs from his travels*. Even in the summer, during the long periods of evening light, the family voluntarily shuts the Persian drapes to keep the heat and the sunlight out. For, unlike their compatriots of the twentieth century, the social milieus of Loti's era do not seek out the summer sun and deliberately keep the abode dark every night. (In the evening: lamps are always left off during the summer and aren't turned back on again until the winter. In order to save money, lights are only used in rooms regularly used by individuals. Most lighting in the house is thus largely derived from the glow of fireplaces). This doesn't prevent the cat from distinguishing forms and sizes.

As a consequence of this heightened desire to keep the inside of the house dimly lit, the cat can feel a profound contrast with the outside: brightness (the L-shaped courtyard, the southern side, "more aery and sunny and more in bloom than most gardens in the city"), a variety of shades of green (plants, bushes, vines), darker in some places (along the common house [*mitoyenne*] on the east side or in the salon full of plants on the west side), lighter in others (sunnier in the center of the courtyard), since cats seem to detect this color in its diverse combinations, which she must also know how to differentiate, by WeAvING, cLInGinG and FUR-vIbRAtiNG, PaWPaddiNg AGAINST the TexTurEs. While she also quiver-SnIFFs:—a mixture of multiple aromas that are variable (from one season to the next) . . . surrounded and invaded by others' scents . . . close by [from the adjacent houses] . . . and neighboring smells (from the fruit trees and prairies) . . . circulating, always arriving from the same side [the westerly wind: the tide coming from the neighboring Charente region, the sea] . . . —

> I want to come back one last time to the writing style I have chosen in this present work. By this style, dear readers, I mean the TYpOGRaPHIcAl games that open the book, permeate throughout, and conclude it as well and which perhaps intrigue you. In the wake of seeking out some form of concrete language, I was inspired by the fact that cats express themselves through gestures. I subsequently attempted to transform these gestures into a figurative language, and I sought to perform this language through a writing style that would be just as figurative, in order to emphasize and approach the living animal. Western abstract languages reserve such practices for literary games, such as those found in the work of Apollinaire and Dos Passos. Other languages cultivate a more commonplace use of such a mode of writing. Most notable here is the Chinese language, which is very concrete, with a wealth of terms, but poor in abstract terms, prior to any European influence that will eventually impose its concepts. Each word expresses a concrete aspect depicted by a drawn character, nuanced through calligraphy. The idea is therefore contained in the form, the contents in the container.[5] Inspired by the example of the Chinese language, but in a rudimentary manner since our Western languages are not accustomed to or adapted to this style, I've merely tried to inscribe movement into words, that is, to inscribe signifying gestures, considered as a whole.

Thanks to the relative permanence of the aromas, Moumoutte is able to establish an olfactory cartography, observe her environment, quickly locate a new scent, and just as quickly activate her curiosity, to be suspicious. She also performs this act of locating by turning an ear to record and situate: —. . . constant, steady noise . . . rippling (the sound of the water fountain in the garden), circulating, blowing [the plants and leaves] . . . whimpering ["*the long moaning of the western wind coming from the sea*"], punctual chirping and snapping sounds [*the chickens in the garden on one side along with the birds and frogs of the surrounding maritime plains*], the whirrings and buzzings in the vast brightness of day [swallows beneath the canopy and insects from spring to summer] . . . but suddenly rubbing and screeching . . . "If anything stirred in the branches of the ivy, she would dart out herself like a young fury, with her fur erect down to her tail, impossible to hold, and no longer her decorous self."[6]

In complete contrast, from its very first years, the house must have appeared dead and gloomy— . . . the clamoring arrival of humans becoming more and more diminished, and then gone completely [a calm street] . . . the sound of creaking wood here and there [from the parquet floors, furniture, and woodwork] . . . the continuous crackling of the fireplace echoing from the chimneys [in the kitchen, during the winter and on summer nights], the sounds of whipping fabric [from voluminous dresses] . . . stifled murmurs [from a servant, or the mother, or the aunt already getting up in years, old and tired] . . . long stretches of silence [*the grandparents constantly knitting*], *Loti writing and reading when he is there*], the ghostly echoes of ringing bells [from the clocks and dinner bells] . . . loud nocturnal sounds [*the reading of Bible verses in the evening all together led by the mother, who began the ritual after the death of her husband*] . . . rare brouhahas [*rare visits*], rare bouts of hissing, rare calamity [*rare mice? Loti doesn't mention them*] . . . —*More than likely, living in a house that is already silent due to the family's preference, where the Calvinist rigor leads one to control one's expressiveness, then "silent as a tomb" after the death of the father, where mourning and duty imposed a sadness onto the family's behaviors, the cat probably clung onto the faint communal acoustics of human sound, in the same manner that she probably was hardly able to grasp the outlines of the dark human silhouettes [all of them dressed in black, perpetually in mourning] from the orchestrated ambient luminosity.*

More than likely, it's by way of scents that Moumoutte is best capable of distinguishing objects and humans—the various attenuated fragrances of rivals [the dogs and former cats] . . . the punctual smells, quickly dispersed, just as quickly fleeting, of objects and beings [food quickly gobbled up, dirty laundry quickly washed, visitors who are quick to depart, mice that are quickly hunted down, newly blossomed flowers] . . . constant pregnant exhalations—humid, woody, scents of melting candle wax, the smell of firewood turned to ash [poorly heated rooms, swollen plaster and wood, old furniture, chimneys], stationary odors [chairs, armchairs, bedrooms and beds] . . . transpired soapy aromas, enveloping [humans and their clothes doubtless washed much more often in this demanding milieu, more than that of today] circulating . . . —which she locates, follows, and certainly maps out, without necessarily emphasizing them more than any other odors.[7]

And she will hardly feel solicited, stimulated in regard to this, whereas her arrival and initial encounters in the new locale involved gallivanting, running and playing, rubbing up against things, deploying all the activities of kittens in order to discover and master their environment, to mark it and construct their territory. For these humans partake in a way of life that is both particular and characteristic of the bourgeois world of the nineteenth century. It was characteristic as well of the calm, ordered life of studious leisure, a fact particularly true in regard to the aunt and mother, already getting up in years to a rather old age for the era—seventy years old and sixty-seven years old respectively in 1880, as noted by the various states of mourning and duties, exhausted from taking care of the bedridden great-aunt Rosalie, whose death at ninety-one years of age in 1880 is a relief but also leaves them alone as Loti travels quite often.

Above all else, each of them lives mostly in their bedrooms, a habit acquired when the house served as a home for four different generations both on the maternal and paternal sides—the impoverished unmarried members as well as the oldest family members all living in familial solidarity—and where the bedroom allowed for a bit of preservation of privacy and the dinner table served as the one place where everyone mixed together or during after dinner collective gatherings in the salon or in the garden during the summer. Each sister sought to keep busy and not be lazy (mostly through sewing, making clothes, and embroidery), whereas other household cleaning duties were taken care of by hired help. Autonomous, individual, silent work favored a more still, sedentary, and contemplative life, interspersed with hushed discussions when they were held in the same place.

Within such a context, they hardly ever intervened into the life of the cat, not calling to her much, even going so far as to use the French polite tense of "vous" when speaking to it. ("Allez-vous-en"),[8] always maintaining a proper distance and never truly naming her: the term "Moumoutte," bequeathed by Loti, was in fact used for all the cats in the house and served as a kind of generic term, similar to the name, "Monsieur Souris" once used by the family for earlier cats or like the name "Jaunet" used by the Fabre family. What's more, the ladies of the house didn't necessarily refer to her by this name but were perhaps more accustomed to referring to her as "mademoiselle" in the same way they referred to male cats as "monsieur," whereas among themselves, they spoke of "this cat."[9]

Rather, it's the cat herself who approaches the humans, soliciting them occasionally, brushing up against the ends of the ample dresses, leaping up, turning, and sitting down on their laps when they are seated, taking advantage of the additional warmth they provide, feeling their bodily vibrations and emanations. Or she climbs up onto the table and settles down right on top of the book they are reading, perhaps in order to elicit a reaction or lays down in a nearby work-basket to seek a warm refuge, or fixates on another nearby object or makes another pass of her tail in order to mark the higher regions of their clothes and face, the lone part of the body discovered with her paws, and exchanges more scents, perhaps creating and maintaining a common odor and thereby integrating them into her world through such a confirmation. *The descriptors used by Loti for the cat* ("little demon," "force," "without discretion," "unexpected," "obstinate," etc.) *denote that she is not invited or called but rather is disruptive of the overarching polite atmosphere as well as the common practice of the era, that of keeping cats at a distance, but she also elicits a pleasure that one mustn't admit too much* . . . , that one must just as quickly control through reprimand (with restraint) or by brandishing the chimney swift (. . . without actually using it) so as to not appear marginally strange in one's engagement with the animal. The cat finds her best responses and acceptance in Clarisse, — . . . pupils: body, gestures, eyes pointed at her . . . ear-turn: addressed acute sounds, repeated, recognized . . . fur-vibration: touches above, on the side . . . then flattening out, twisting herself, purring . . . also like pushing her paws behind her . . . or to join her: the quiver-sniff from a distance, following the scent that becomes stronger as she goes, raising her paws up high [to the highest floor of the house], penetrating, strong aromas [the bedroom] . . . —there inside she will share with her human not so much an abundance of interactions but a copresence comprised of aromas, attention, listening, silence, mixed with sleep; the cat thereby transforms the area into one of her privileged zones.[10]

Such a relation is not by chance. One can connect it back to the attention that Clarisse had devoted to many other cats for quite some time, helping her sister Nadine, integrating the familial abode for such a contact with cats and remaining unmarried. We must also emphasize the strong similarity of the atmosphere, attitudes, and interactions with the cat to those of Loti's own childhood experience growing up; there is a striking parallel due to the bourgeois culture he resided in where one wouldn't think of abandoning or transforming this cultural attitude for the sake of the cat; rather one would adapt and apply this culture onto the cat, with the result of forcing her to evolve within a very particular environment that she will adapt to. As a child, Julien Viaud was shaped to live a fairly autonomous life without too much interference and he duly frightened his great-aunts, who always wondered what was wrong with him when he was around them. Physical interactions were not commonplace and the gentlest of interactions for him came in the form of sitting down on the laps of these women. He was almost never referred to by his first name, following the rules of the local school where this was judged to be improper and where one's family name was used, and so at home he was referred to by the generic moniker of "petit" in a similar fashion as the cats were referred to by generic names like "Mimi" for male cats or "Moumoutte" for females. And among the adults, he was simply referred to as "the child." Clarisse was much more his caretaker than any of the others, playing with the boy every now and then in her bedroom or on a Sunday afternoon, where he would spend long hours observing and studying things and the shadows that filled the room. He also did this as well spending time with his great-aunts and grandparents![11]

Such a rapprochement shows that during this time period one does no more for an animal than one does for a child. But one can treat them in a similar fashion and the cat therefore ends up benefiting from a kind of familial integration, albeit in accordance with a social milieu of a specific era. The goal behind such an incorporation is to benefit from Moumoutte, since she can make things happen: *speak to her a bit, pet her on occasion, above all watch her* evolve, *contemplate her* "unexpected awakenings" when she LeAPs, along with her "reflections" when she fixates, MOTIONLESS, on her "curious ideas" if she suddenly juMPs *and subsequently enjoy oneself in watching this* "little devil" *during a lazy afternoon or during "a quiet evening." This form of company, which has recently been shown to help attenuate negative emotions, is plenty of interaction for them.* In the same way that the cat is fine being considered as "a toy" (which Westerners will stop making use of in the twentieth century considering that such an adequation is not sufficient for the interaction between domestic animals and humans but which shows precisely the relativity and malleability of the contents hidden behind the so-called permanence of words and concepts), an attitude that should suffice for this cat since during this era a pampered animal is basically one that is fed and housed.

In this regard, it is a good bet that the hired help, heralding from the countryside of Oléron, the ancestral region of the Viaud family, solicits the cat even less than the others except when it is time to provide her with her daily feeding. She is given meat, still a rare occurrence during this era for house cats, since a good number of milieus still consume meat in small amounts and instead provide bread, even chestnuts, potatoes, and cooked carrots. Furthermore, a diet of meat for felines is still poorly considered in the available treatises concerning cats inside bourgeois culture in the middle of the nineteenth century, since meat was considered to make them "voracious and carnivorous."[12] No doubt, *but Loti makes no mention of it*, the cat knows how to identify this human, to the extent that she must learn to regularly approach her in order to be fed, even solicit her attention at certain times, always at the same time of day which the cat must learn, *always in the morning, suggested to the maid in order to prevent the cats from endlessly begging for food, since their physiology incites them to eat often but in small quantities and will lead them to cry out for more if one gives in to their demands*. The cat is often near her since the maid is always present, *even when the sisters are absent*, but the maid appears to only be understood as one element among many in Moumoutte's environment, but obviously an important one, without entirely being detached from her, and learns to adapt to the *cultural distance* maintained by the rural maid. Moumoutte thus enjoys an autonomous life, perhaps symbolized by her fairly meager deployment of meowing (*Loti evokes more so exchanges of glances, but this could perhaps merely be a question of the author's penchant for contemplation*).

Once an adult, Moumoutte concretizes her autonomy through long sedentary stays inside this now largely known and mastered territory, which, however, is seldom interactive. During the cold (of winter) she lingers close to the warmth (the chimney), s-p-r-e-a-d o-u-t on softness (carpet) or "CuRLeD uP" in a ball on an even warmer area (the armchair), obviously in the copresence of humans whom she must follow into their preferred environments, as the other environments will remain cold.

Humans who allow for this influence over objects that elsewhere would be considered exaggerated. In the heat (of summer) first and foremost outside (the courtyard), she will willingly take leave of staying close to *humans, who often prefer to remain* in the shade inside the house or under the tree canopy: "In the summer she would have the languors of a Creole. During entire days she would lie in a stupor of comfort and of heat, crouched on the old walls among the honeysuckle and the roses, or stretched on the ground, presenting to the burning sun her white stomach on the white stones between the pots of blooming cactus." Moumoutte must fill up such sedentary time with activities: contemplate or lazily bathe in relaxation, no longer detached from alimentary concerns; slumbering here and there: a frequent occurrence when lacking a proper stimulus, but bearing a whole procession of dreams, *which have henceforth become objects of study for several other species as well.* Or she may become bored, a situation that had also been suspected to be experienced by animals but which older documentations rarely make mention of.[13]

> Earlier on, I spoke of the themes, angles, and concepts in the humans sciences that should be mobilized in respect to the study of animals in order to enrich our readings of them. So as to demonstrate their utility, I'm gathering together here what they allow us to see in regards to this cat, Moumoutte. She perceives and feels this world better and better by way of doing, that is, by moving and acting, thereby entangling actions-perceptions-representations: each acting upon the other, which she then maps out, organizing her environment, with which she co-determines and co-develops. She transforms it, remains within it, and above all acts in particular areas, dispersing gestures, sounds, and odors, constructing her territory with specific preferences she has established through her daily movements, brief stops, and actions according to a given moment. She transforms herself through contact, refining her perceptions, her representations, acquiring forms of knowledges, forms of behavior, adopting and expressing attitudes, postures that function as signs sent to others around her.
>
> She perceives others as both an element within her environment and also as a particularity since they act, and she acquires understanding of this through doing: moving, feeling, representing to herself, interacting. She codetermines her-

self just as much with them, constructing forms of knowledge and behavior according to specific moments, places, and gestures of each one of the others and their encounters, according to the modalities of copresence, contact, and interaction, organizing her territory with specific privileged areas of coexistence or separation, adapting her attitudes according to each being, shaping shared preferences, allocating her time, moments of pause, and actions accordingly. She co-develops with them, slowly but surely fortifying a taste for reciprocal presence, brief interactions, parallel manners, instants of autonomous activities, contemplation, and studious leisure, each one of them adjusting themselves and reinforcing each other within this common path.[14]

And yet, if the cultural sciences help us to see that our various ways of being are constructed, biological sciences remind us that these ways of being are also given by emphasizing the specificity of the animal in question, the animal's possibilities and ways of perceiving, feeling, and representing to itself, acting according to biological affiliation, in other words: according to what it is capable of doing. Tying these two approaches together allows us to see that this given influences the constructed, which in its own turn then modulates the given and on and on. Here, the given and the constructed are not in contradiction but reinforce each other since the environment confirms this and therefore consolidates, even accentuates the genetic or epigenetic capital. Indeed, the Angora is purportedly lazy, indolent, and sleepy. This notoriety is not merely the result of human exaggeration but is also derived from frequent behavioral observations, since these behaviors were retained and concentrated in this specific breed through human selection. They are exacerbated by a number of pet owners who seek out these animals and expect such traits, demanding them, and even encourage such behavior by the lifestyle they propose to the cat, à la Loti. Such behaviors are developed by cats placed within this environment and they slowly but surely structure themselves accordingly.

The example of Moumoutte Blanche living in the house and alongside the Viaud family highlights two aspects that, during the specific time period, favored the development of this kind of cat and that make it characteristic of a specific social milieu and era and subsequently its social diffusion and perpetuation up until today in Western civilization: the mode of the Angora, often sought out by those who wanted a cat for their salon; the major role played by women within the widespread acceptance of the companion species of the cat, most notably in bourgeois society, adapted to the modes and lifestyles similar to those of the Viaud family.[15] However, other cats, confronted with similar environments, would construct themselves in the same way, and many men such as Loti take comfort in this model since it corresponds to their idea of a cat.

Just like the refined ladies of the house, Loti first and foremost enjoys Moumoutte through contemplating her, by way of intercepting her gaze, which he subsequently translates into a story in an anthropomorphic manner through words, privileging as a human and intellectual, vision and words to the detriment of sounds, gestures, and odors, convinced as he is that cats also carefully examine men. Persuaded that men should also examine cats, Loti is one of the first writers to provide a documentation of the human desire and quest to understand the animal point of view: "Through the strange windows of her eyes, the task will be to penetrate deep into the depths of unknowable brain hidden behind them." Faced with this impossibility, Loti wants at least to ensure Moumoutte's autonomy. As is the case with a great number of other humans, this freedom only reinforces his conviction to intervene as little as possible, which only reinforces Moumoutte's behavior.

He reproduces the same situation he experienced as a child when his grandparents, great-aunts, and aunts observed him evolve around them, without intervening or talking much to him. Above all else, he forbids himself from interacting with Moumoutte based on his own conceptions of cats: intelligent, elegant creatures, and aristocratic and therefore proud and irritable, "snobbishly distant" and contemplative. A representation strongly forged from the very fact that cats were placed into such environments, constructing themselves as a function of them, acting by conforming to these particular human environments and therefore . . . justifying this so-called conception of them frequent within bourgeois and intellectual milieus. Furthermore, this connection between human desire and animal reality has been recently emphasized once again in regard to contemporary humans and cats.

No doubt, Moumoutte more than likely notices the human more when she perceives his form, scent, voice, becoming more lively and more mobile (beginning in 1884). *With Loti*, she feels and sees herself snatched up by her hindquarters, raised up and *placed on him, gently stroked and pet-*

ted, which reassures her and leads her to accepting such shackles whereas she had not originally solicited it. She more than likely considers him to be particular within these regular moments together. To such an extent that she quickly grasps that she can enter into areas in the house where aromas arise (the dining room) when she sees he is present, sitting down next to him, receiving a morsel of food, and gently begging him: "*occasionally reminding me* of her presence with a discreet tap of her paw across *my dinner napkin*, and sneaking mouthfuls of food[. . .] from the tip of *my fork*." She will acquire a knowledge of an interaction and a specific moment (feeding), which will continue on to our current times as a privileged and essential site (in the metaphorical sense) when not entirely exclusive, for contacts, complicities, concordances through gestures between most domestic cats and their *owners*. Conversely, she almost never is solicited to interact in other ways, nor does she herself do the soliciting; hardly ever hearing her name *called*, she does not *call out* from her side. Moreover, the fact that a human and a cat mutually recognize each other's voice appears to be unique to our current era.[16]

> I want to reiterate once again what the human sciences allow us to see. This cat perhaps doesn't simply engage in living through interactions with Pierre Loti (he doesn't mention it) but lives out a much stronger copresence with the very dwelling itself. This forces Moumoutte to decipher the home, to know it, and understand it. In the same way as the hired help, she must slowly but surely connect their gestural encounters to certain goals (eating, going, sleeping, returning), to certain spaces (kitchen, courtyard, wine cellar . . .), to certain moments (meals, evenings, mornings . . .), and therefore connect her representation of it all to this human—connecting his actions with hers. For Moumoutte is restructuring herself and adopting adapted behaviors: obey, precede, warn, solicit, and even avoid. She undergoes a reciprocal configuration with the common collective daily activities, complementary gestures, and shared connivances. She lives through another process when Loti vanishes and then returns. Each time he reappears, she must learn—and ever the more quickly—to mobilize her memories and her prior acquired skills, to recover and adjust to the exchanges, reconstructing their complicity.

Moumoutte must also adapt to the new forms and immobile scents (a myriad of objects brought back from Loti's voyages to Asia starting in 1884, which Loti parcels out into various rooms and in the garden), new sounds heard regularly (the piano), thunderous noise at certain times (construction and rearrangement of various rooms due to renters arriving beginning 1884, about whom we don't know very much at all, but who remained there for eighteen years due to financial reasons), acrid, persistent odors (the scent of paint, glue, and wallpaper) in certain places that will completely turn her territory upside down, forcing Moumoutte to reconfigure it, to no longer move through it the same way, and to stop doing so for a time in order to observe and once again get a feel for it. In the same way she must also become accustomed to the waves and flows of cries, scents, and agitations (*during each reception upon the completion of another one of Loti's exotic rooms starting in 1884*). *None of this is mentioned in Loti's book, The Book of Pity and of Death*, given that Loti clearly wanted to unify his narrative by retaining a calm atmosphere from earlier times. However, in his journals, we indeed find more reliable accounts regarding such calamities.

Among these agitated, chirping beings over which hovered a strong stench, recognized as much by their faces that the cat knows well, Moumoutte no doubt takes notice of more pronounced presences: a changing procession (by the *successive orders given by Loti the naval officer), until only one remains (Pierre Scoarnec, a twenty-three-year-old young man from Brittany kept on as hired help beginning in 1886). Then there is a punctual presence (Loti's niece, who regularly comes and goes), or sudden appearances or more permanent but calmer arrivals (Loti's wife in 1886 and two servants in 1887).*[17] *The sadness that grips the household in 1887, when one of Loti's children dies during childbirth, and the subsequent depression that grips his wife more than likely didn't increase any interactions between the humans and Moumoutte. Perhaps with the exception of Pierre Scoarnec.*

Certainly, Moumoutte must have quickly perceived Scoarnec's sudden arrival and shuffling around the home, which bears a new face along with new emitted sounds (the laughter of this gambler and boaster). Perhaps Moumoutte also follows his path (as is recounted by Loti when he writes of Scoarnec descending the staircase on a Monday Easter morning in 1886, leading Loti to note that he is awakening the house). There is also a new atmosphere that slowly emanates throughout the home due to

Scoarnec's arrival (he is considered likeable, kind, cheerful, and thoughtful; Scoarnec gets along quite well with Loti's mother, aunt, and the maid whom he reinvigorates while quickly integrating into the family). And Moumoutte is quick to detect such atmospheric changes through the vibrations, intonations, scents, and perhaps even through the pheromones diffused throughout the house, which cats are well versed in detecting since they as well communicate through such means. She also experiences a disruption in her daily life as a result: Moumoutte can no longer stay inside at night, finding herself dragged along *with Scoarnec* each night to a new area to sleep *beside him* (*the cave that serves as the wine cellar at the end of the courtyard that runs up against Scoarnec's bedroom*). Moumoutte quickly adapts to this, learning to detect the same repetitive sounds (*Loti's mother signaling to Pierre when his workday is done and it's bedtime*) that precede another departure. "At the very first words of the sentence, even though they were pronounced in a low tone of voice, White Moumoutte cocked her ear anxiously. Then, when she was convinced that she had heard aright, she jumped down from her chair, and with an air at once important and agitated, she ran by herself to the door, in order to go in front, and to do so on foot, never allowing herself to be carried—wishing to enter into her bed-chamber of her own free-will or not at all."[18]

She also interacts with . . . Suleïma—the turtle—brought back by Loti from Algeria in 1869, who begins hibernating every November in a box and only reemerges in the month of March to live out in the courtyard, between the sunlit cobbles and the flowerbeds. She always rediscovers the turtle "with the same new sense of surprise," always carefully observing Suleïma initially and leaping on her when she sees the turtle moving the potted flowers, pawing at her carapace, recognizes the uselessness of the act, gives up, and eventually is content with merely following her around and finally ends up napping alongside her beneath the radiating warmth (under the summer sun). Moumoutte simply tolerates this animal, merely engaging in the feline capacity which traverses species borders, while hunting and chasing the others living around her, including her own feline peers (a number of whom roam about in the store entryways and on nearby rooftops), attacking them as soon as they appear on the garden walls, always wanting to go out when she suddenly hears the other cats' meowing outside, leaping up onto the windowsill when she sees another cat on the other side, her feline solitude leading her to produce a strong territorial sentiment, in defense of her well-being, which we now know is not necessarily obligatory in this species but which can be activated at the available genetic level and further developed and maintained in relation to the environmental situation and individual character of the cat.

Conversely, she will also head out for encounters elsewhere outside her territory, even at night, climbing up in the tree branches and onto medium-sized walls, passing through gardens and across rooftops. *Normally, the family doesn't authorize such nightly wanderings*. But Moumoutte, "like so many other" neighboring cats, holds fast to certain principles that are nonnegotiable." *Such nightly escapades provoke scandalous reactions among the ladies of the house, as does* Moumoutte's refusal to go to sleep in the cave in the courtyard. *But she oscillates between the rigor of an "austere" and bourgeois milieu, applied just as much to children who aren't allowed to hang out in the streets, and her desires to be roaming. This is in contrast to many other cats who are only allowed to gather together during chosen and controlled visits, prepared in advance and observed by their human owners, along with the unspoken conviction that one can't control this aspect of cat behavior (castration is still uncommon and only reserved for males). So it is that Moumoutte oscillates between a desire to be educated and an ineluctable renunciation of such a proposition. Between a specific desire for a culture and a nature judged to be unsurpassable. And all the while, in the background, hovers a representation of the cat as a natural, wild being. A creature with instinctive behaviors and an innate, permanent temperament and thus the cat is consequently considered more a guest than a member of the family.*

From the very first years of her stay chez Loti, Moumoutte regularly becomes pregnant and lactates. And upon each occasion, she quickly sees her kittens taken from her—*the hired help must have drowned them as was the custom of the era*—while only being allowed to keep one of them. In this way, *the family applied a certain control over the births of the cats in the same way it did in regard to the number of their own children in the bourgeois society of the era*. Moumoutte is allowed to feed and educate her one kitten until the time of weaning and then loses her just as quickly, as she is *gifted or inserted into the local or familial exchange network from which Moumoutte came*. Her activities, research, choice, gestation, pregnancy, lactating, education, *about which Loti hardly writes a word*, take a great deal of time and energy, periodically distracting her from any time spent with humans, which incites her to distance herself from them, maintaining this distance and thereby reinforcing her autonomy.[19]

And in this way, Moumoutte Blanche will experience a test with the arrival of Moumoutte Chinoise (picked up in China, at Chefoo harbor in the summer of 1885), who remained at sea for a long time with Loti (in Loti's ship cabin) and who will enter into the environment of the house (March 7, 1886) at the age of one year old. For a while, this second cat will live alone in another closed off place (*inside the old art studio of Loti's sister, which will later be transformed by Loti into a Gothic styled room in 1887*). She will still be able to discover and assimilate the odors, sounds, colors, and humans who visit her, wanting to acclimate her to dissuade her from running away and allowing the necessary time for Moumoutte Blanche to recover from a difficult birth *leading to the death of her kittens*. Moumoutte Blanche must have made ear-turns, quiver-sniffing the intruder, getting emotional around this invisible presence and she attacks Moumoutte Chinoise as soon as she catches a glimpse of her once freed as she walks into the kitchen—puPIL-DiVerTING-DILATING raiSES PAWS fuR TAIL HEAD EARS back ROUNDED s,p',I,T' DE-PUpillates l-a-u-n-c-h-es p-a-w-s LEAPs CLAWS "de-claws" rEDDenS MIXEs It UP CIRCLING me-ows . . . — smells, grasps, draws forth, displaced, swept up, perched; "seated in a corner, pensive and somber," observing with "her big eyes." Moumoutte Blanche no doubt perceives, through the emanations, forms, and colors, that this is not a neighbor's cat, that it is connected to one of her surrounding humans (Loti), who takes her in his arms, slowly relaxing her, thus demonstrating a typical gesture and releasing an odor incorporating the human for marking and mixing purposes. Moreover, she most certainly had already detected the new aroma, the common scent of this human, once she climbed on him and rubbed against him. And this is probably why Moumoutte Blanche doesn't attack Moumoutte Chinoise when she is placed back on the ground but merely walks past her without looking at her, feigning ignorance. During the early days, the new cat is left to wander around the premises. Moumoutte Blanche follows her and watches her endlessly explore the courtyard and her other favorite spots, her vegetation, since Moumoutte Chinoise had only ever lived on a Chinese sailing ship and then a French naval vessel.

In the end, Moumoutte Blanche chooses reconciliation for, during a moment of clarity, "she approached with deliberation, and, then, all of a sudden, she sniffed straight in the face of the other one" close to the glands that disperse the pheromone F4, which aids in cats rubbing against each other in order to reduce aggressivity. This substance subsequently provokes both a mutual sniffing and rubbing, allowing for an exchange of scents signifying a desire for cohabitation creating, as long as it's repeated, an identical and particular odor. The cats become inseparable and go everywhere together, use the same area to go to the bathroom, and even eat off the same plate, roaming and both getting pregnant at the same time and both lactating indifferently, each of them allowed to keep and nurse one kitten. Moumoutte Blanche even teaches Moumoutte Chinoise how to beg at mealtime, to languish on the cobblestones in the courtyard under the summer sun, to settle in on the rugs close to the fireplace during the winter months, and to climb onto the armchairs, where they will roll around and nap together, keeping warm.[20]

> There are certainly a number of similar "anecdotes" on the social capacities of cats, your cats. Today, they have become accepted and taken into account by ethologists, after having for a long time been laughed at and rejected. In this case, the older cat has inculcated her ways of living to the younger cat, integrating her into the older cat's culture. This process follows precisely what the human sciences emphasize in such an exchange, namely what they refer to as "entanglement, enmeshed," which one must add to the other social processes of socialization and cooperation only recently evoked by ethologists. What takes place is a process of recomposition of self through the modification of attitudes and sociability, the reduction of distance, the acceptance of proximity, the elaboration of a common life and therefore a subsequent creation of codes and rules where priority is given to adjustment, complicity, and tolerance.

However, Moumoutte Chinoise follows more closely *the gestures and actions of humans*, expressing a superior need for proximity and protection, soliciting more tactile contact with Loti or Moumoutte Blanche, is more interested in the courtyard—even staying out there well into winter—endlessly exploring the leaves and tending not to stray off so much. Whereas Moumoutte Blanche is more indolent, more autonomous, more anxious when in close proximity—even with Moumoutte Chinoise, whom she often rejects, preferring to head to their sleeping quarters in the wine cellar by herself in the evening, whereas Moumoutte Chinoise is often carried there, *and this is precisely what we see in the photograph of the cats with Loti.* Moumoutte Blanche also remains more territorial, still chasing off other cats, tending to run off more, and continues to retain her privileged relation with Aunt Clarisse, leaving Moumoutte Chinoise with Loti, which maintains the distinction between the two: *the aunt seeming to be more distant than Loti.* These behavioral differences—whether created or maintained—highlight the role played by breeds or lineages, that is, genetics, or perhaps even epigenetics, as well as individual characteristics, education, and the initial structurations (in regard to their personalities). However, all these aspects do not impose themselves in a mechanical, unavoidable, and definitive manner. Rather, they are mixed with other environmental factors and social-cultural constructions within an ongoing dynamic. Feline ethologists perceive this when they discern behavioral modifications that take hold after a cat's initial development.[21]

Both cats end up dying around the same time in the summer of 1889 from the same infection *according to Loti, who doesn't provide us with the exact name of the illness but gives us enough details to know* it's contagious since another cat had just trespassed into their territory right before. *One of those feline illnesses that one wasn't able to treat, which was poorly discerned during that era and which spread so quickly according to the veterinarians that they had a difficult time studying it until the middle of the twentieth century. According to Loti's diary,* Moumoutte Blanche *was the first cat to die and not the second, as he describes it in his novel, Lives of Two Cats. He more than likely structured the novel in this manner in order to begin and end writing of Moumoutte Blanche in a way that preserved cohesive unity.*

There are three aspects revealing the extent to which Loti's family privileged Moumoutte Blanche. First, they make calls to veterinarians—a rare practice for cats during that era—all the more so given that veterinarians at that time didn't know much about cats, who were thought so little of that an animal doctor would not spend time learning about them, and so a mastery over cat examinations and illnesses was not common. One would still have to wait until the end of the century for a veterinary specialization in cats to take hold . . . in Paris. And during this period, the veterinarians are unable to provide any cures, which leads the Viaud family to seek out popular remedies that are just as useless. In contrast to the common tendency to quickly kill off the cat in order to avoid the spread of the disease and contamination, the family remains by the cat's side all the way until the end—

giving it a place in the writer's room; a novelty which had begun in affluent bourgeois households, following the image of Zola and his dogs. In the end, Moumoutte Blanche is not thrown out with the trash like most dead cats—an end that Loti dreads for the second cat, who flees in order to die, horrifying Loti, but who is nevertheless provided a resting place in the garden in order to symbolically reinforce their connection.[22]

The case of Moumoutte Blanche shows that the behavior of humans among themselves along with their representation of felines structures their attitudes toward their cats, and that this human environment influences a cat's behavior through a phenomenon of reciprocity: their attitude confirms the convictions and attitudes of humans, which reinforces those of felines. As a result, these humans do nothing exceptional (no brutality or games, for example) in order to truly exit their environment, to truly stand out, except to feed and pet the cats, which above all else simply leads the cat to form an attachment—in reality more to the place than to the person. It's as if the cat considered humans as merely elements within her territory and not as separate or detached from it, in the foreground, but along with the rest of the elements, as if the territory took precedence over them. Her behavior coheres with that of the common Angora cat of the salon, since these individuals have an analogous genetic heritage and live in similar environments.

> A similar process is at work in similar environments but perhaps a bit more closed off in the case of interior bourgeois environments. The latter tend to lead to ways of behaving that are somewhat accentuated in relation to other common housecats, which situates such cats in contrast to wild or stray cats, these . . .

CHAPTER 3

Inside Cats

. . . which more than likely represent the most frequent example of the majority of pet cats in the nineteenth century, since this type of feline seems to have been largely developed in aristocratic residences as well as urban bourgeois lodging. And here again, the details we have concerning these cats come from writers. Culturally, these writers are part of the bourgeoisie, but their position as writers bestows upon them a particular status, which they use as a means of transmission that must be taken into consideration. At the same time, they are the ones who largely elaborate, and quite frankly formulate, the bourgeois representation of cats, which leads to proposing an environment for cats that is specifically calibrated for the singular traits of cats living inside.

Bibiche, *encountered by Michelet during a stay at a friend's house in Geneva in 1868, can never leave her living space due to Michelet's fear that something might happen to her outside, as an earlier cat had been poisoned, no doubt due to one of the baited traps freely dispersed throughout the cities during this time to limit the number of feral cats and dogs. Or perhaps she had eaten a poisoned mouse.* Bibiche divides her time between contemplation, sleeping, and moving about her living area—which she knows quite well—that doubtless hardly ever changes and probably provides little stimulation except for when visitors stop by. Seated at "her favorite window," she observes stray cats walking across the rooftop of a neighboring building, "sometimes emitting tiny particular meows," then turns her head, more than likely split between a dissatisfaction, notably during periods of reproduction, a desire to join them, and a certain suspicion of the unknown creatures. Perhaps so as to fend off boredom and create interactions, she turns toward her humans, solicits (she doesn't wait) being petted, even being played with. Every now and then, *especially around dinner time*, when she must smell the aromas of meals and the gathering of *humans—often more available during these brief moments—* "she seeks out an old ball," "brings it to her owners," emits "impassioned, expressive meows, imperative even, as if to force their attention toward her," using this meow as an instrument for interspecies communication, which appears far from being frequent. This plea for attention, which appears as play, is so rare during this era—*humans rarely having fun with their animals, just as they don't play with their children—that Athénaïs, who didn't like this "captive toy" and who preferred house cats that were left to freely roam, believes such* a gesture expresses nothing more than a desire for kittens, a situation of a nervous pregnancy.[1]

This partial return toward humans is also, around 1850, due to the Angora cat raised by Charles Dickens, born in his house and raised in his study. The cat follows the writer everywhere, perhaps due to being deaf (a genetic abnormality common in white Angoras bred with each other in order to preserve their precious color, which led to strong consanguinity), which forces him to master his environment through sight. He sits right next to Dickens while the author reads and writes and solicits being petted, prepared to stomp out the lit candle with a paw in order to get Dickens's attention. All this is limited due to the cat's infirmity, which prevents oral communication; *moreover the family doesn't even give the cat a name, never directly addressing it. Neither does Dickens, who merely contents himself with the cat's presence.*[2]

Démonette, a black Angora (thus from a crossbred lineage) given to *Barbey d'Aurevilly at the age of one month by the wife of a doctor in 1884*, hardly ever leaves the study and never the apartment. The cat thus constantly marks the apartment, transforming it into her territory, as she will do, *following the death of the writer*, in another place, "replete with cat odor." Whereas she hides from other humans and is quite suspicious of them, the cat will construct rare interactions with *the writer and his governess. And his governess will seek the attention of Démonette as well. Bar-*

bey d'Aurevilly often invites the governess to eat dinner with him, which the cat must learn to predict based off aromas, lighting, and noise: she climbs up on the neighboring chair, props her paws up on the table, and partakes in the morsels of food that *the writer offers her*, giving a slight tap to her human when asking for more. She grows accustomed to meowing briefly to the governess when the latter leans toward her, always adopting the same timbre, exclaiming the same phrase: "Do you love me?" which the cat can recognize by the sameness of the sounds, thus deducing the interpellation and affective intention of this corporeal dispositive, similar to one adopted for speaking to breastfeeding babies, which we have since proven to work with cats as well as dogs in order to better understand such solicitations. *But during this time period, such an interaction is so rare that a certain Octave Mirbeau, having paid a visit to his colleague, and who is also an animal lover, is "struck by* this way of responding."[3]

For, most inside cats are quite far from having turned toward interacting with humans in such a manner. Such as Mouton, judged to be a "magnificent Angora." And yet he is black and white. The vulgarization of this breed has made it difficult to retain an Angora with only white fur, and thus many people have come to define an Angora as bearing a long chest coat of fur. "He came from a good bourgeois family, not averse to the wondrous effects of an ample amount of fur beneath the velvet pillows," thus, a milieu where cats were sought for their appearance and not for their relation. Gifted to Jules Michelet and placed in his apartment, Mouton oscillates between feeble solicitations and sleeping. In the cloistered space and perhaps due to an epigenetic capital, Mouton quickly adopts the habit of taking "long naps steeped in a leisurely reverie." Mouton retains this way of being *when Jules, in marrying Athénaïs, moves into a house in Neuilly*. The cat rarely leaves the house (to such a degree that he eventually dies by gunshot, unaccustomed to being suspicious), not even following *Athénaïs who lives on the second floor*, preferring instead to stay on the ground floor near the wood stove in the kitchen and the pantry, idly spending his days largely sleeping inside this house. The daily rituals of his humans don't stimulate Mouton much at all: *Jules, writing in the morning and departing in the afternoon while Athénaïs reads, sews, and cleans the home*. So it is that Mouton takes to a "long and feeble existence of servitude and inaction," leading *others to proclaim* that he "is nothing more than a thing," whereas Minette, a street cat, *initiated to such an indolent life by Michelet in his Parisian apartment*, becomes analogous to

Moumoutte Blanche. Two divergent paths, which also lead us to suppose an endless play between genetics, environment, and personality.[4]

> These ways of being, between a return toward humans and that of an independent withdrawal, are connected to each individual, in a particular, and therefore original way by . . .

Pierrot: In Good Company

(Paris-Neuilly, circa 1855–1862)

> . . . whose portrait is sketched by Théophile Gautier in his novel, *Ménagerie intime* (1869), a novel that evokes various animals near and dear to Gautier throughout time, following the same literary model as *L'Histoire de mes bêtes* by Alexandre Dumas (1867). We have emphasized on several occasions the entanglement of literary description and literary invention that led to doubts as to the actual existence of these animals.[1] Such muddling is unavoidable in literature. Such practices don't simply lead to pure literary invention as the actual existence of Pierrot is attested to elsewhere. Quite the contrary: it can actually help to structure reality since the daily lives of both the humans and animals are influenced by literary representations. And the imprecisions surrounding the depiction of Pierrot—Gautier was not very rigorous in regard to the chronological and geographical aspects of the cat's history, completely erasing any mention of the cat's move to Neuilly within an entirely different narrative framework used to unify his story within a more unique tone akin to that written by Loti—can all be restrained, bypassed, and rectified through recourse to other documentation. In this case, it's not recourse to Gautier's other correspondence but the publication of two of the three books of memories by his older sister, Judith Gautier, who doesn't write much about Pierrot but nevertheless certifies his presence and is however, much more prolific when mentioning the familial framework, people, and the atmosphere comprising the environment of the animal, with and in which he constructs his ways of being.

We know nothing of the birth or motherly education of this "white Angora," *given to Gautier*, "still just a baby," more than likely severed from his mother, *by the actions of a young artist according to the story presented by Gautier, or by the artist's matron if we follow the comments made by Judith Gautier, around the year 1855. Quickly after his reception, he is named* Pierrot *due to his white color, then* Don Pierrot de Navarre, *due to the nobility attributed to his fur and poses. The use of the first name, even* Don Pierrot, *appears to be quickly accepted among the adults when referring to him. However, the writings we have at our disposal do not mention what name is used when the animal is addressed directly or if the latter—similar to cats in the Fabre family and the Loti family—merely hears the term "cat" or "minet" when being spoken to.*

The bourgeoisie of the era residing in apartments seem to land somewhere in the middle. If the feline is addressed with the term "minet," it can be understood as a term of endearment or affection, in the manner of Renan and his wife.[2] However, the name remains an interchangeable one for cats, thought to be themselves just as interchangeable because they are judged to be identical. To address a cat with a specific name, as Barbey d'Aurevilly and his governess do with Démonette, indicates a supplementary degree of insertion into the family and a recognition of individuality, even of personality. In actuality, not worrying about addressing or speaking to the cat, or simply using the name "cat" or "minet" or merely a name used among the other humans but without addressing the cat, or even calling to the cat by the name "cat" or "minet" or even using an actually individual name given to the cat—often some kind of diminutive—to speak to him, all cloak various conceptions, practices, and different human gestures. These do not have the same meaning for the cat in question, who doesn't see, hear, or feel in the same way as the humans and who can react in a different manner.

And yet, it would appear that the use of a specific name for addressing a cat began to develop starting during this period, with the upswing in inside cats, as if their particular interior environment, of confinement and proximity, was favorable to such a shift. But while this living framework is the necessary condition for this trend, it alone is not enough. Humans must also be prepared to speak in such a manner in order for their cats to respond, which seems to be the case. This practice is subsequently fortified and diffused, little by little, within geographical or social space, as well as within time, culminating in its common occurrence in our current societies.[3] It is quite possible that such a use of a specific first name for a cat had already become commonplace in the Gautier household, since Théophile opens his writing with thoughts regarding Pierrot's personality, situated at the highest level of his memory in regard to their cohabitation.

Little by little, Pierrot will master this space, first seen as vast, then slowly carved out into a smaller place following the cat's own growth and development. The space will become compartmentalized based on the surrounding obstacles (Judith often suggests closing the doors for fear of the cat escaping and in order to preserve the quietude of her relatives in the fairly constrained space), the people living there, the various gestures, smells, noise, and luminosity.—In the shaded side of the home [next to the courtyard]: noises shocks endless sliding [kitchen], different alternating voices [the Alsatian maid and cook], permanent aromas [furniture, contraptions, burnt wood or coal], the changing odors of various mixtures

[food], constant warmth; darker areas [a long and narrow room], obstacles used for orientation and observation (*beds, armoires, desks*], permanently maintained scents that are similar but different [girls], precipitated movements and noises [*Judith is ten years old upon the cat's arrival and she is spontaneous and lively*] or slower movements [*Estelle is seven years old and has a much calmer and discreet manner*]. On the brighter side of the home [*on the side closest to the street and terrace*]: another saturation of sensations [the mother's bedroom], a calm, motionless human [attending to *reading and work*] present for long periods, then absent [*travels*]; another scent that is distinct but just as permanent and dense [*Théophile's room, the lone male*] of another human who is calm and motionless (*writing*) or noisily stirring [*getting himself ready, singing, reciting*] or who vanishes [*work, travels, and the demands of daily life*]; and a more luminous area [*the salon-dining room area between the relatives' bedroom and three French windows*], fixed aromas [of the furniture], the comings-and-goings [*of humans*], the brief stays of others [*guests*], exterior sounds, agitations, frequent passages, noises, myriad instants of shouting, arriving, passing, disappearing [*family meals, various parties, piano recitals*], intermittent silences, warmth *[the principal fireplace*] . . . —

According to Théophile, the cat becomes accustomed to the bustling life of these humans, *in particular to the worldly life of a writer, to the constant visits and discussion in the salon*; the cat doesn't go hide or flee from such occasions, quite the contrary: Pierrot accompanies Théophile, observes, listens, solicits. "He participated in the life of the household [. . .]. Seated in his normal place, near the fire, he truly seemed to understand the conversations and take interest in them. He followed the eyes of the interlocutors, letting out a cry here and there."[4] Pierrot reacted to each of these environments, atmospheres, temporalities, instants, taking on certain poses, making certain gestures, emitting sounds that represent as much reactions to the *human signs* he receives as they do signs sent out to other living beings, *which seem to react in turn. As Gautier himself suggests: to take an interest in the cat—to respond and solicit Pierrot's attention*—leads to the formation of a symbolic interaction between them, that is, communication, by way of a particular language.

> I see the police of the animal-human distinction are beginning to become annoyed. I can hear them fulminating: Animals do not possess language. They merely have communication!

Linguists have indeed theorized as such. They have naively defined language . . . according to the human (a capacity to express a thought and to communicate it through language) and they claimed, just as naively, that language doesn't exist as such . . . in other species. The latter would therefore only have communication at their disposal: the means for exchanging information by way of signs.

Such an anthropocentrism that makes the capacities of the living be defined by humanity—amusing by its naivety, irritating due to its pride—is incredibly embedded inside us, Westerners, such that this reasoning seems normal, obvious, and natural to us. In reality, it rests on the fact that we continue to circulate a two-thousand-year-old conception constructed by the Greek city-states, which we claim as the first creators of democracy but which were ethnocentric and built around hierarchies, affirming strong differences between men and women, the free and those enslaved, the Greeks and the barbarians . . . humans and animals.[5] Theorized by philosophers from antiquity, reprised by the majoritarian version of Christianity, transmitted to philosophy and Western sciences, this conception imagines the animal world in the form of a pyramid, with insects deemed to be the most rudimentary followed by animals estimated as being more and more complex (but not too complex, and their numbers are becoming smaller and smaller!), from one step up to the next until one finally made it to the top of the summit . . . man, proclaimed as being the most complete and accomplished living being, and as such, the very measure of the world. It thus seemed logical to define capacities of the living from atop his position on the pyramid and then examine if such capacities also existed further down below. But of course, this view was never once logical, given that a horse, a dog, or a cat are not . . . humans! Conflating human intelligence, reason, consciousness, etc., with the definition of Intelligence, Reason, Consciousness, and therefore armed with human definitions of such capacities, we have been able to confirm up until today that animals do not possess intelligence, reason, consciousness, language, etc.[6]

This pyramidal conception has never been proven and is merely a philosophical construction that has henceforth become cultural, even ideological and political, since it is used to justify human interests and their pretensions concerning animals. This conception has very recently been rejected by science, which has proposed another schema: branching. Starting from the last universal common ancestor, species, and therefore humans as well, have evolved in diverse ways, in all directions, and groups within the common branches then develop individually on specific twigs or stems. With such a schema, another schema is established! It becomes absurd to think that other species are armed with human definitions of capacities, whereas the human species is merely one branch among others. Branching must undo hierarchy, without denying the specificity of the human, each species possessing their own uniqueness, and must interest itself not so much in what other species don't possess in relation to the human but rather in the originality and richness of each species. We must therefore de-anthroposize concepts as I have mentioned in regard to that of territory. It becomes logical and necessary to show and elevate definitions to the level of abstraction so as to finally free them from their human version, to forge more general meanings and specify them for each species. This has been done for a number of physical aspects. As such, it no longer is a surprise to hear the statement that "plants breathe," since the definition of respiration has been elevated to the level of chemical exchanges. But we are much more reticent to do so with regard to mental capacities, since such a shift in thinking them would erase the hierarchical conception of the activity that elevates the human over *the animal*. But this is, however, the path one must take and that ethologists make use of without theorizing it. They have done so in regard to intelligence by defining it as the aptitude to adapt to one's changing environment. Such shift in definitions has allowed for the recognition that a great many other species are also intelligent . . . each in their own way, that there is not simply one form of intelligence (defined according to the human) that would be judged superior over the others, but rather a plurality of intelligences exist (according to each species).

The de-anthroposization of the notion of language is under way. Semioticians have been able to define it as the capacity to communicate through signs (be they emotional, vocal, gestural, tactile, or olfactive . . .) comprising a system resulting from an acquisition. And yet, zoo-semioticians have noted that this is also the case in regard to relations in animal so-

> cieties or human and animal relations. And zoo-sociologists have demonstrated that these latter relations can generate symbolic interactions, and that such interactions do not have any sort of obligation for human articulated language in order to function, that is, that they possess their own bodily, gestural, emotional, and sonic language, more than likely vibrational, olfactory, and pheromonal that humans can barely detect at all, even though they also emit it.[7] But let's get back to Pierrot.

He "takes on a charming amiability." *Théophile believes that he acts* "like all animals which one takes care of and spoils." However, other coddled felines who are churlish when it comes to novelty, such as Démonette, show that there must be more to it than mere coddling and spoiling them. Pierrot forges himself and engages in the trials of novelty, and then in the very habit of the imposed novelty and sociability. More than likely, he also possesses an adapted personality (contemporaries consider the Angora as intelligent and sociable . . . if he is solicited), which is not rejected, blunted, or weakened by the environment but rather is constructed such that he can reinforce it, confirm it, and, to the contrary, fully deploy it within this environment in order to construct a personality. In fact, Pierrot becomes affable and sociable just like . . . *Théophile*! He lives one of those interspecies influences in cohabitation that ethologists have begun to study and indicate for dogs and now contemporary cats as well, which gives credit to the old popular adage: *like master, like dog*. As such, he makes the living room, *the heart of the apartment*, the center of his territory, drawn to the warmth of the fireplace, the light, the possibility of imprinting different scents, noises, and movements, which he then transforms through his presence, his markings, his own odor of a un-neutered male cat (which Théophile's contemporaries note as being a very strong odor, but which he doesn't mention at all, perhaps out of pudeur or habit, or perhaps because Angoras or at least this specific one would smell less than others), as well as the interactions he is able to solicit or incite.[8]

More than likely there is not much interaction with the maid, who is also from a rural milieu foreign to this cat of the salon, even if the cat will learn from her how to sit in the corner, "close to the fire," and no doubt also to perform her "bathroom business" in its ashes, *as was im-*

posed on cats in order to not make a mess in the rest of the house like we do today, with our litter boxes. Pierrot certainly had more interactions with the cook, whom he must have quickly distinguished from the others by connecting her with the kitchen and pantry area, *which is where she more than likely spent most of her time except when she left to go grocery shopping.* Pierrot more than likely sat close to her, begging and rubbing against her legs, scratching and meowing, hoping for a gift of milk or a small morsel of food—*the latter having already been a common gift to the cats by Théophile's family members. We don't know* whether Pierrot begged for food, negotiated, or if he merely waited and ate what was given to him without a fuss, which is more than likely, given his personality. He more than likely became *accustomed to the regular hours of snacks probably imposed by the cook, who had better things to do than to be bothered by the cat and who was forced to watch over him,* which also led to a kind of discipline facilitated by his affable personality, which merely strengthened it. Thus fed, Pierrot no longer feels the need, nor apparently the desire, to leave the house at night to seek out other sources of food elsewhere. Henceforth well fed, he can reduce the expanse of his vital living area, as many contemporary cats do today; with this idle relation attached to the reception of food, cats can then dedicate themselves to other activities: most notably turning toward *idle humans.*[9]

We don't know anything about the Pierrot's relationship with Mme. Gauthier, which Théophile never mentions; Judith describes that time as being divided between the worldly life of her husband—his travels, notably a long stay in Nice from November 1856 to April 1857—and time spent in studious leisure in her bedroom, which was also a veritable living room. We can hypothesize the relationship that Pierrot had with Théophile's two daughters, thanks not to anything written by Théophile but rather to the writings again of Judith. When the cat arrives in 1855, the two sisters are hardly busy with anything, Théophile having neglected their studies. They spend quite a lot of time playing with the cat when the writer decides to finally take them under his wing, in 1856, providing them with a domestic tutor. Judith notes that their mother writes to them from Nice "telling them to leave the cat alone," indicating that they must be treating Pierrot like a doll, clothing him, putting him in disguises for their numerous short plays in this quasi theatre that structures the life of their beloved parents.

I want to return here to all this information in order to invite you to place yourself beside Pierrot, to attempt to approach what he must be living and experiencing in such moments:

— . . . EAR-TURN, repeated sounds repeated approACHING, INvitING . . . "Pupil-Turn" STATIONARY masses now gROWING in the distance . . . Pupil-SpreADING . . . LOwering-HEad EArs WHiskers TAil [submission] FLATtens himself grasped raised up carried off . . . or he s-l-o-w-l-y pushes out his paws . . . turns over, gets up, sits, lays down, gets up, sits, turns over, flattens out gets up lays down . . . rubbing up against coarse material with back head paws [clothes] . . . ShAkINg hIs fUr WrITHinG CLinGing . . . sc-am-pe-ring HIDinG . . . — *quickly tiring of the games that hardly test his abilities, in contrast to what is extolled today*, only remaining interested and resigned, patient, in order to glean several caresses and kisses here and there, which lead him, no doubt, to accept being handled but not too much![10]

His relations with Théophile are better known, given the writer inserts anecdotes about them in the middle of his writings, but without mentioning their eventual particularity. Pierrot seems to enter into the bedroom of this human at will, a space saturated with various scents from this human, who sleeps and works long hours therein. Such a situation allows for the cat to assimilate odors and ways of being, no doubt including sounds and gestures, bringing together perception and representation, including that of Théophile. The cat is thus able to perceive a candid and profound state of his human, since the scents denote the writer's true affective, physical, emotional, and overall state of health. Or these scents detected by Pierrot on his human may also denote the immediate travels and encounters of the moment, *whereas the human is only capable of doing so by way of perceiving faces, postures, and discourses, often dissimulating something, and which can therefore mislead, thereby creating a fair amount of uncertainty in their interpretation.*

In a parallel manner, Pierrot constructs his perception-representation of this space, assimilating it and making it his territory, traversing it and marking it, thereby receiving the right to climb around, in particular on the desk, which allows him to better observe his surroundings in order to protect himself and to subsequently procure a better sense of his well-being, an effect that has recently been proven in the behavior of contemporary cats. He can also occupy himself with books (that is, he is granted this access thanks to his amiability, *which is granted to him due to the good opinion of cats held by those surrounding him*), sleep on top of them, look at the pages, even turn them "with his claws." Or even bother an *amused Théophile*: as soon as "I pick up the pen, he immediately leaps up on the reading desk and posits a profound stare of attention at the steeled beak of writer scrawling down ink on a field of paper, turning his head following the slightest gesture at each beginning of a new line. Sometimes, he even attempts to take the pen from out of our hand." Whereas Théophile rejoices in what he affirms, with much humor and distance, *as the passion of a writerly cat*, who also needed to fall asleep every now and then on top of certain books, "as if he had [. . .] read a specific trendy novel," Pierrot partakes in the feline interest of stirring, quick-moving objects: virtual prey that he follows adeptly with his quick, jerky movements of the eyes, and these subtle sounds of disturbance of the turning of book pages and the scribbling of ink on paper are perhaps the only sounds echoing in the bedroom, stimulating him and leading him to be on the lookout and eventually claw at them. *Even if their readings differ, Gautier and Pierrot nevertheless construct an interaction that brings them together, a quasi game, even though Théophile hardly plays games with his daughters and the humans who created games to play with cats during this era are few and far between.*

Such a comradery only continues to grow between the two, as Pierrot will spend many hours curled up in this human's lap. There, on his lap, he discovers the ideal temperature for a cat (80–95 degrees F) while also benefiting from Théophile's working hours, morning to night, thanks to a weak light source and a low level of sound that they both appreciate, which reassures them and which must have relaxed Pierrot all the more. He can certainly feel the vibrations and emanations arising from the human and can subsequently examine them, controlling them in his own way, perhaps only climbing up when he feels Théophile is willing, thus adapting to his world, having already integrated the writer into his own territory by marking him. But is the cat seeking nothing more than physiological and sensorial well-being, or is he not also behaving in such a way as to express and signify an attachment, making use of his biological aptitudes while also overcoming them?

Moreover, Pierrot seems to cumulate a desire, a need, an affect, a communication when he modifies his purr in order to express his refusal to climb down off the lap of Théophile, expressing his desire and will to stay curled up in the lap of the writer, creating a sonic variation *that Théophile can hear, humans possessing the capacity to distinguish sounds of a varying intensity . . . if they so choose, and understand, leaving the cat to do as he pleases.* A close contact of Gautier, a connoisseur of the writer's own various states of being and attached to him, Pierrot will learn to manifest his own capacity to vary his expressivity, inventing a form of communication and adapting himself. And it's possible the cat also achieved such an expressive variation with his sight. "Sometimes, when he was sitting in front of you, he looks at you with such sweet, soft and cuddly human eyes."

> One could make the case that perhaps this was nothing more than literary invention, human projection, an anthropomorphism, since many among us don't believe in the communicative capacity of the animal gaze, which would only be reserved for humans. Nevertheless, I still wonder whether or not this isn't a process similar to the social adaptation we see cats undergo through scent, purring, meowing, rubbing, and thus a feline reality that Théophile is not making up but is quick to read in his own, obviously anthropomorphic, way. If this was in fact the case—but we are still in the realm of hypotheses, since the animal gaze has hardly been studied until now and the feline gaze has been reduced to the mere fluttering of the eyes as a mere amicable sign—we must then envision two possibilities that do not exclude each other. The first is that such a gaze is already practiced among cats, as we have recently discovered is the case with horses and dogs capable of emitting ocular signs of emotion. And then, it has only recently been paid attention to by humans starting in the nineteenth century. Or, like Loti's Moumouttes, where the human also refers to their gaze and also interprets it in the same way as Gautier, and certainly other cats who would be placed within similar environmental conditions, Pierrot would have slowly but surely grasped the importance of his gaze in relation to his humans, in particular for these contemplative writers or the women living within their bourgeois condition of

> the era. To the extent that cats can discern various faces, they must also pay attention to these other details. Pierrot would have perhaps also learned to mobilize this means of transmitting expectations and emotions through his gaze.[11]

In the evening, Pierrot also expresses an attachment when he waits for Théophile's return at the front entryway to the house. Pierrot has acquired the ability to differentiate between the members of the family, registering their presence, paying attention to when they leave or are absent, and retaining a memory of their regular routines of returning home. Pierrot thus imposes an organized, repeated, and sequential gestural presence, *which Gautier notices and accepts, taking pleasure in participating with him, delighted to be of importance to the cat*: Pierrot brushes up against his legs, in order to renew belonging, rounds his back in order to receive a reciprocal rubbing, purrs, in order to manifest his well-being, heads toward Gautier's bedroom, "waiting until we undressed and then he hopped up into bed with us, gathering our neck between his paws, pushing up against our nose with his, licking us with his tiny little pink tongue, rough like a nail file, while letting out quiet, inarticulate cries, expressing in the clearest of fashions his satisfaction for seeing us return. Then, once these tender exchanges had settled down and the time for sleep arrived, he would perch himself on the back of the smaller bed and fall asleep in perfect harmony, like a bird on a tree branch. As soon *as we awoke*, he would return and stretch out beside us until we finally got out of bed."

As such, head against head, according to the privileged mode of how cats identify with other cats, Pierrot renews the exchange of scents in order to fortify his feeling of community, applying facial pheromones (notably F3 for belonging and F4 for tolerance, as we have already seen) by rubbing with his mouth and forehead, as felines are known to do with objects resting at the same level as their head, logically using the means at his disposal within the sensorial framework of his species. However, he also will add other means, such as these soft sounds, perhaps in response to *Gautier who doesn't indicate what he is doing to the cat—if he is petting him, speaking to him, or imitating him through onomatopoeia*. And here again, he modifies a feline function, adjusting it in order to grant it additional dimensions: emotional, social, and cultural dimensions (because they are constructed) with interspecies communication in mind. In the same way Pierrot adjusts his initiative of interspecies contact at the proper moment—between *laying down and sleeping, awaking and getting up*, which he had carefully taken notice of, perhaps being comforted and encourage by the voice of *Théophile calling out to him. It would therefore seem that Théophile was one of the rare individuals of the era capable of deciphering felines in their own way. As an individual and a human, in a way that was more than likely suitable for them since he demonstrated that humans (if they want!) can grasp the vocalized meaning—not of cats in general—but of their specific cat*, the cat himself adapting his vocalizations to their relationship. Pierrot thus creates a regulated and ritual interaction with Théophile, perhaps comprising a supplementary dimension.

Such a tale could appear somewhat far-fetched if we merely remained at the level of the writer's interpretation of events. During evening parties when Théophile would be away from home, Pierrot would wait for his return at the entryway until midnight—the typical hour of the writer's return. On several occasions having waited there until 2:00 a.m. for a tardy Gautier, Pierrot would give up and simply go to sleep. *Théophile quickly analyzes this act on the part of Pierrot as a kind of "silent protest" since "our behavior doesn't please him." Saddened by the rupture, Gautier makes an effort to always return home by midnight in order to restore the ritual between himself and Pierrot*, "who once again took up his nocturnal post in the anti-chamber."

> At the risk of disappointing you, it's not a useful notion to posit something like a feline mastery of clock time and believe in something like moral reprobation, as Gautier does. However, this doesn't mean that the story isn't true if we put ourselves in the place of the cat and take into account his abilities. Among animals, and the cat in particular, there are indeed other ways of mastering time. We have difficulty thinking and seeing these other forms of temporal mastery since we struggle to get outside of our human condition, being merely humans and not absolute spirits. And, as we have erected the human form of measure as the measure of the world, we believe that the absence of our ways in others signifies, by way of anthropocentrism, that they simply have nothing to offer, or that other animal relations are the same as ours, due to an erroneous anthropocentrism.

Pierrot could have simply made use of typical feline perceptions in order to slowly piece together the hour when Gautier would reappear—by locating at that moment the level of nocturnal luminosity or noticing a reduction of *Théophile's scent in the area, since the odor would have certainly changed from the beginning of his departure, until a threshold of scent was reached that the writer would again stabilize and signal by way of returning home at the same time every night and which would then increase again thanks to his presence.* Luminous or aromatic intensities perceived and mastered by cats much better than humans would have allowed Pierrot to construct a representation of his human. So, he would have given up waiting by denoting the discord between his measurements and the late return of the writer's scent. And there would only be a new harmony once Gautier and his scent had returned to the entryway at midnight over several evenings.[12]

> And yet, there is also another possibility: Pierrot develops, displays, and imposes what sociologists refer to as a performative behavior. Indeed, he does things with his body, which leads Gautier to subsequently respond to them (who certainly wants to be attentive to his cat), and Pierrot thus creates his world through adjustment, negotiation, and reciprocal configurations. This can only be possible through the combination of several factors that become interconnected and dynamic: an adapted genetic—perhaps epigenetic—capital; conditions of lodging that favor relations; an early and continuous socialization of the cat, endlessly activating the social dimension of his brain, allowing him to assimilate the given norms and develop appropriate responses to them; a predisposed attachment to humans maintained and reinforced through the responses made by the cat, himself becoming more and more confident and apt; and finally, an epigenetics that is slowly constructed, imprinted, and reinforces all this.[13]

"And this was a cat who was very well-loved," even though male cats were considered to be less affectionate, less attached to humans than female cats since many of the males could leave the house and would wander out great distances, changing environments, in contrast to Pierrot, who was forged and raised in a different way. And yet, Pierrot's turn toward humans is contained, limited by *Théophile's perceptions, which are very similar to those of Loti: that a cat is contemplative, tranquil, tidy, independent, therefore one shouldn't bother him too much and force him to do something; that the cat accepts becoming a companion of presence and grants his affection, which is already difficult to obtain, while not asking for anything else. But the writer is more daring than many others such as Mallarmé or Catulle Mendés, who consider an apartment cat as some kind of "ornament" "set atop a piece of furniture" that one admires for the cat's beauty but whom they hardly consider as affectionate and whose attention they almost never solicit, the remnants of a relation established during the Ancien Régime. Nevertheless, Pierrot is often left by himself due to Théophile's and his wife's absences as well as the increasing time the tutor spends with their daughters.*[14]

Does he spend a lot of his time looking out windows? Does he paw here and there to occupy himself? Does he become fat for lack of activity as is the case for many cats in our contemporary era? *We don't know*. He seems to sleep a lot, compensating for his lack of activity and stimulation in an environment that is all too constrained, too familiar, and seeks out the warm areas by the fireplace and on the laps of humans in order to counter the loss of heat as his body temperature lowers while sleeping. He more than likely remains in a light sleep (which is 70 percent of the time for contemporary cats) in order to react quickly if need be or simply to register his surroundings from a good observation point such as the salon.

While cats are dreaded and not allowed in many residences—the number of domestic caged birds is quite high during this time period, and although Pierrot doesn't seem to spy on them all the time, "there are a number of canaries a friend gave from an aviary and never returned to pick up which sing out loudly, the one trying to outdo the other, filling the apartment with strident rolls"—his personality and the conditions that place him in front of them attenuate his desire to hunt, a fact that has been measured in current cats as well. Furthermore, he also doesn't pay attention to the meowing coming from outside, high above (from the rooftop), which he hears since Théophile can as well; Pierrot doesn't even leave his lap since he can't see his peers and can't go join them (*in an apartment on the fifth floor*), whereas the strays can't come and beg. Pierrot responds even more than Moumoutte *to the urban bourgeois desire to not let him* wander, become thin, dirty, "flea-infested," irascible, in order to have *instead an apathetic lazybones who submits all the better to their wishes, so as to apply to their cats and dogs what is already put into practice for children, in this case for Gautier's daughters, who hardly ever leave the apartment anymore.*

Pierrot will also get along with the family's Bengali cat. And when Théophile's daughter's tutor moves in for a brief period while his wife is away between November 1856 and April 1857, Pierrot also engages in peaceful relations with the tutor's tabby cat, who also moves in with the family: "Don Pierrot blinks his sweet eyes in order to welcome his guest," an affirmation that no one will immediately label as anthropomorphic and set aside but which we now know can be understood as a friendly sign, confirming that some (and not all) humans from the era could read their cats well, an ability that was more than likely reinforced due to the close quarters in which they all resided and the empathy of the cat owners. Conversely, the tabby cat, perhaps heralding from the streets, raised alone in an apartment, doesn't see things in the same way, illustrating the fact that isolated individuals don't know how to read others, leading them to panic when traveling or in a new situation, as is the case here. "In crawling out of the basket, he was curled up and completely frightened [. . .], he spit on Pierrot's nose, and, as soon as he's able to flee, in a prodigious leap, he hops on top of the armoire and disappears behind the rosewood pediment. And there, he more than likely observes, to take in his new surroundings at leisure." Climbing back down, he approaches Pierrot, licks him several times—a sign of a desire to insert himself within the group, a symbolic interaction that happens as much in the presence of other cats as it does with humans, in order to allow for a social life. *And that's all we know* except that Pierrot doesn't reject him, and here as well affirms flexibility and a gentle demeanor and behavior that was not created but reinforced, developed by the conditions of life, within an incessant co-influence between the individual and his environment.[15]

Pierrot experiences a rupture during *the family's move to a house in Neuilly in April 1857*. Enclosed in a basket, disturbed by the new odors and sounds, by the streets, gardens, and parks, by the impediment of his wired-basket visions and the shaking experienced, he expressed "his anxiety by way of several disapproving meows." He discovers the world of the garden, life with a female cat *adopted by his humans as soon as they arrived,* subsequently leading to reproduction, but does not become a hunter or a wanderer. He will even become quite friendly with a couple of white Norwegian rats, purchased and placed inside the home in a big cage, and eventually he will become acquainted with their offspring. Even when the rats are let out, he will respect their space, scratching and meowing at the door to the room where they are held, even when it's closed, spending long hours next to the cage, watching them play, while other animals will come sleep alongside him thus constructing an interspecies friendship for which there are numerous other examples, from birds to dogs, requiring the young animals to quickly become accustomed to each other, or requiring adapted personalities as is the case here. Such an interest in the rats shows that Pierrot prefers being inside the house rather than in the garden, like Mouton with the Michelet family, once he has mastered, recognized, and transformed the territories into a new territory, as long as he is able to carve out a familiar life, albeit restructured, *given the daughters now leave every day to attend school, and Théophile now lives in Paris, with receptions in the evening, and his wife is now present and active in the bedroom.*

The cat's preference for staying inside the house could be *due to the humidity of the site, located near the Seine with its tendency toward flooding,* which Pierrot would have had difficulty dealing with. And even inside the home, *which is more humid and more difficult to keep warm than the former apartment, which also bothers Théophile who eventually installs, in 1859, two heating apparatuses—one in the vestibule, the other in his work area, where, from now on, the temperature never drops lower than 72 degrees F—"we felt a bit smothered and no one wanted to spend time in there with him," which says quite a bit about the temperatures in the other rooms of the house.* Only Pierrot, "who needs the atmosphere of a hothouse," will remain, resting on his lap for hours. And it will be precisely in this very work area that he will eventually die, at an undisclosed date, no later than 1862 it would seem, "after a year with a cough" caused by a "cold which would quickly degenerate into consumption, actually thoracic tuberculosis, the symptoms of which he presents (anemia, weakening, continuous feeling of being cold that leads him to seek out warmth . . .) perhaps transmitted to him by way of a contaminated piece of cow lung or from a fellow cat who lived in an unhygienic milieu, *and against which a "doctor" (more than likely a veterinarian often referred to as such during this time period) can do nothing. The family buries Pierrot at the back of the garden beneath a white rose bush.*[16]

The call for a veterinarian and the cat's burial could lead one to think that their manner of tending to their pet's ill-health or demise is identical to our contemporary culture and has persisted up until today. In truth, appearances hide a rather significant evolution: such practices were still quite rare at this time, where cats were almost never medically attended to and their cadavers were most often simply tossed out, whereas today their burial has become commonplace with the accumulation of wealth of humans and the growing normalization of cats as pets. In history, the frequency of an event is much more important than its presence or absence alone, which if we take it into account, often gives the impression of a flat and unchanging history since the dynamics at work have been erased. As such, the different lives, manners, and cultures we have just explored have gone through numerous variations of intensity and statistical shifts in one direction or another up until our present, as we shall see a bit later.

Moreover, contrasts can be discerned between each of these territorial cultures, according to the position and importance of the humans in the environment: outside and weak as regards stray cats, peripheral and marginal for vagabonds; more central and pronounced for inside cats, who have a less primal territorial attachment due to a limited engagement with it. These dynamics give rise to various ways of being, cultures, and thus to different cats, such as Zola depicted in a short story. At the turn of the twentieth century, we also have the example of the writer François Coppé who lived with several cats: one Angora in the salon, a second common black cat, who liked to hunt, fight, and chase after female cats, and a third stray street cat who wanted nothing to do with the others, "barely even looking at them," the gap being too large between their various biocultural ways of living.[17] The differences between cats during this period were such that a number of humans believed in actual separate and distinct feline natures, even conceiving of the possibility of different species, as some humans would even posit in regard to human social classes, in an era where there were whispers and postulates circulating regarding this same idea for various human "breeds." However, a small cadre of people did observe the behavioral plasticity at work in cats and maintain . . .

PART 2

Possibles: One Culture Arising out of Another

. . . even if they neglected or couldn't see the various transformations themselves: pets who become vagabonds (who run away or are abandoned), vagabonds who become strays, which can even happen to adult cats through obligation, environmental changes, and constraints. As Gaston Percheron notes, "If we completely abandon a cat," it just so happens that he will "run away from the house and seek out the forest where he will live a life out in the wild. He will then become indifferent toward humans when he doesn't simply try to completely avoid them." The disinterest or ignorance of the writers in such scenarios comes from the absence of humans within the transformational processes of cats. Such aspects are not dealt with here but they still remain up until our contemporary times. The transformation of common cats into domestic pets provide us with some evidence. A writer by the name of Rédarès, author of one of the first French treatises on the education of the cat, addressed to an affluent urban female readership desiring company, proposes a dark portrait painted by Buffon regarding common cats. Nevertheless, he affirms that with patience "a bit of time and good principles, one can transform them into loyal and obeying subjects." He does note certain cats remain resistant to this domestication: some "species" tend to be "undisciplinable" (probably referring to strays, and you'll notice that he "biologizes" categories that are really biocultural) and adult cats and males are more difficult to deal with than younger cats and females, an opinion still shared by contemporary ethologists. However, such an opinion has more to do with a perceived frequency than a behavior that one could consider as systematic. As for the rest of the cats, Rédarès focuses on "instruction," understanding that it is a question of social processes whose possible and positive results he summarizes within the human environment of the era: "When cats have formed social habits, they will retain them until their death; well-raised cats are sober, sweet, and calm. They are not tempted by pillaging or theft, they even procure a pleasure in playfully catching mice and rats, fully occupied with pleasing their owner."[1] These . . .

CHAPTER 4

Conversions

. . . tend to be punctual and individual. They display constructions, deconstructions, and reconstructions of cultures and ways of being and therefore denote the possible adaptation of cats and the role and influence of the environment, since such processes depend on both human and feline initiatives and thus particular circumstances for both humans and felines. Rédarès proposes an active conversion, guided by humans, with daily "brief exercises" to get them to stand straight, offer up a paw, keep their claws withdrawn, remain still lying down, and run or leap after a piece of paper. Such advice is still heeded today in order to train one's cat, but during Rédarès's era such an approach was quite rare. However, documents appear to show a British precedent. In his poem "Jubilate Agno," which was composed sometime between 1759 and 1763 but remained in manuscript form and unknown until its publication in 1939, Christopher Smart, a friend of Samuel Johnson, well known for his privileged relation with felines, paints a portrait of his own cat, Jeoffry, and explains that he can sit with a look of seriousness, go and fetch and bring back objects, leap over a baton, capture a wine cork, and let it be tossed again, etc. This leads to an acceptable companionship: Jeoffry doesn't destroy anything as long as he is well fed, shows himself to be thankful, purring when one exclaims that he is a "good cat" (which, moreover, will be used more often than his name). Jeoffry is also friendly with children: proof that such behavior is not a given. Rédarès confirms this for France: the (common) cat must be educated at an early age and, he will add, must remain enclosed in the same favorable environment, which is absolutely necessary but not enough. And he ends by stating that such an environment is rare, and more present in the "larger world" and "high bourgeois society."[1] The conversion of . . .

Trim: A Rat Hunter Turned Companion
(From the *Reliance* to the *Cumberland*, 1797–1804)

... is inscribed within the British context and provides us the occasion to focus our research on the other side of the channel. Trim is born of a rat-hunting cat on a sailing vessel, dedicated to protecting hull and cargo from rodents, a role that is rarely mentioned in texts but which was an ancient and fundamental one since it led to the maritime diffusion of *Felis catus* within these initial dwellings of domestication.[1] Trim is evoked in a manuscript by Matthew Flinders (1774–1884)—an officer of the Royal Navy and explorer of the Australian coastlines—entitled *A Biographical Tribute to the Memory of Trim*, written in 1884 in response to the profound "loss" of Trim.[2] The manuscript was mixed together with a number of other naval papers handed over to National Maritime Museum of Greenwich and all but forgotten until it was rediscovered and published in 1977. In a similar manner as the texts about Jeoffry, the book ended up making Trim famous in the anglophone world since these two cats now resonated with the contemporary ways in which humans and cats got along with each other. The depictions and interactions described between cats and humans in the book have become quite common but were still very marginal during the era the texts were written. Furthermore, Flinders endlessly mentions the exceptional nature of Trim, which appears to be the very origin of the writing, which he assures us is not a mere tale or allegory but the recounting of actual events. If he attributes moral and human values to a cat, no doubt in order to justify an animal biography—a genre of writing nascent at the time in Great Britain and France—he does so with humor and makes it easy for us to detect them. But what is of interest to us here is the following particularity: the feline conversion from one culture to another (which we will refer to as transculturation), which interests us because it takes place within a closed world, wherein both humans and the environment are circumscribed.[3]

This document must obviously be cross-referenced with others, most notably with a document chronicling the death of the naval officer and above all with his travelogue, which was written during the same time period.[4] Not so much as to know more about Trim, who is not mentioned therein, but to have a better idea of the environmental context, such as the times when Flinders is present and absent, or the dates, which are not well documented in the biography. The omission of dates was done in order to both synthesize and unify the tale without giving out any information on the Royal Navy. In effect, the navy officer writes his homage to his cat while a prisoner of the French on the Isle de France, now referred to as Mauritius. He changes the name of his sailing vessels and even the birth year of the

> cat, indicating that he was born in 1799 during a provision run from the African coast to Australia (when in reality he only made one such trip in 1797), no doubt in order to not reveal the exact date of the trip and scramble the reality of English activity in Australia.

Trim is thus born during this journey on the *"Reliance," sent by the governor of Port Jackson (in Sydney harbor) to seek out cattle on the Cape of Good Hope, passing by Cape Horn on the initial voyage and returning by way of the Indian Ocean so as to still benefit from the constant west-east winds. During the return voyage, the sailing vessel is pummeled by such high winds and waves that the captain fears that the ship will capsize at any moment. Such fears are even more heightened due to the age of the vessel, which had already endured a trip from England to Australia in 1795 and is taking on water, leading the ship to have to slow down, prolonging the journey that began on April 11, 1797, until the twenty-sixth of June. Flinders does not mention* the first days of the voyage, nor the first weeks of life of the new male kitten, *whose existence he probably hadn't yet even noticed,* but who must have learned how to experience the environment through the awakening of his senses.

First making use of his thermic, olfactory, and tactile senses in order to quickly locate the mother, to approach her and take hold of a nipple, then to get a feeling for his surroundings:—fur-vibrations shimmering cold water [parcels and packages stored high above, the seeping of water from below] . . . quiver-sniffing a salty humidity marine smells aromas sweat stink heat moisture [kittens, 109 cows, 107 sheep, 3 horses, the ship's crew] . . . coarse cushions [wood, hay] . . . —Then visual and auditory senses; perhaps, like Toto, acquiring these senses more rapidly in such an harsh universe than other cats; the darkness of steerage between decks must have favored the early opening of the cat's eyes: — . . . endless agitations, around masses that continually come and go [men performing maneuvers] . . . a perpetual and strident swarm crackling [waves, wind, hull, and mast] repeated whistling cries and shouting, both long and short, [humans at work] . . . poor orientation . . . pupil-dilated, ear-turned . . . slowly getting oriented . . . —Finally, the ability of locomotion, first fast, then stumbling:— . . . sliiiddding, to one side CLAWing Flattening himself . . . sliiiidding from the other side . . . flattening himSElF clAwINg . . . shaking FaLLing Re-organizing . . . rOLLing In A Ball standing BacK UP More than likely he must regain his balance, then bolt off faster and better than most other terrestrial cats. Throughout his first weeks of life, he must have become immersed in this environment, making it his own, considering it as normal and natural, and began mastering it.

He more than likely received a complete maternal education given he wasn't separated from her, *since the sailors would not interfere with this stage in order for the cats to survive*, letting them take care of themselves, in order to be subsequently more self-sufficient. Unassisted, the mother must have left at times in order to seek out food, perhaps obligating the kittens to learn how to look, listen, and move at a very early age, in the same way she must have severed her ties with them very early on, earlier than today's average of four weeks. This included showing them very quickly how to hunt, since kittens learn through imitation, while simultaneously inculcating them with a certain alimentary taste that will provoke them to want to hunt the same sorts of prey.[5]

Besides making his own attempts, Trim also prolongs this training by no doubt observing other adult felines aboard the ship who are more efficient at this practice. And above all, he learns through playing with his siblings, an essential aspect in order to make him more sociable, cooperative, to have improved agility in body, mind, and the senses. More specifically, cats hone and accelerate their sensory acuity at dusk, a privileged time for them. Trim and the others play on the ship's bridge, at which point, *according to Flinders, he takes notice of them, no doubt during the beginning of the crew's rest hours, the "Reliance" now safely in the harbor of Port Jackson*. They play together like kittens still do today between the age of one or two months, but perhaps these kittens begin to do so a bit earlier: cats severed from their mother early on begin to play together early on. Trim would have been around one month old when the ship arrived into the port. He now has time to discover, know, and master his environment in complete tranquility *since the "Reliance," having arrived at the port in a dilapidated state taking on water, will remain at shore throughout the winter of 1798, for significant repairs. Flinders, now promoted to lieutenant and therefore no longer able to participate on subsequent expeditions around Australia, will still have plenty of time to keep an eye on Trim.*

Within such a milieu and with his good feline education, Trim could have easily become a rat-hunting cat like the others, along with his siblings, but he quickly differentiated himself. He shows himself to be dynamic, mastering speed that contemporary cats only achieve, on average, at six to seven weeks, but without yet possessing the visual acuity which, today, is usually acquired at three months of age. As such, he still poorly evaluates distances, missing targets such as the ship railing— . . . GaLLoPs . . . body t-h-i-nn-ed o-u-t tail s-t-r-et-ched Fur folded back, Ears LOWered . . . leaping climBING holding on, swept away, stumbling, straightening himself, FaLLs . . . FriGHTENED eARS RaISED fur standing on end claws OUT . . . CRAshes, sinks, rises back UP, moving about in the water . . . —; he doesn't lose consciousness, knows to keep his head above water, a skill mastered today between four and six weeks old, and he doesn't drown: "he learned to swim and not be afraid of the water," having become accustomed to getting wet during the storms throughout the ship's voyage to port. *Flinders overhears* that Trim has fallen from the boat and swam more than once, *more than likely safely returned to the boat thanks to one of the sailors making repairs to the ship* in the harbor and then one day, "when we toss him out a rope, he grabs hold of it like a man and climbs back up onto the boat like a cat," having perhaps learned to climb by playing with the rolled up cables or ropes on the ship's bridge— . . . MoVEs PaWs h-e-a-d-dr-a-w-n . . . sEES sMELLS roPE . . . CLAWS paws in front pulls claws paws behind PusHES . . . CLAWS CLAWS HOLDS ON . . . WORKING FUriOUSly . . . tired, DAZED, reassured, shouTING, laughter, suRROUnding by GeStUreS[6] . . . —

Trim's exploits makes him *well known to the ship's crew, who have the time and serenity to take interest in him, to the extent that he expresses his comprehension of human manners: in this case the most important human gesture for sailors and men at sea*. He makes himself all the more *noticed, remarked, and remembered* for his physiognomy, which is *deemed quite particular*: he sports a rather long tail and a tiny, round head, a fairly ample whisker mustache, along with curved ears, all *considered as amiable thus bestowing on him "the most beautiful form ever seen,"* and he also sports black fur, which would make his appearance common, but with white patches on his paws and stomach and chest in the shape of a star, *which makes him easily recognizable, making one believe his is predestined: "It's as if nature had deemed him as a model and prince of his breed."*

Busy attending to the construction work on the ship from the forecastle deck, the officers see Trim become accustomed to laying down in front of them, "right in the middle of their walkway" in the position of a "lion at rest," which we might refer to today as the posture of the Sphinx, which forces them to stop and admire him, "contemplate his amiable form and beautiful white paws," to speak and laugh about his attitude while the cat is right next to them, persuading them that he is "excessively proud of himself," reading his posture as a sort of "beautiful vanity" but without the pretension common in many humans, provoking neither jealously nor anger according to Flinders, but rather a touching affection, we will say, which prepares the men for particular relations with the cat and reinforces them. And these men don't take notice of just any old cat, for there are a great many others on the boat, but rather this specific cat, who places himself in their line of sight, soliciting their attention as well.

You've just read the human side of the story. I will now return to the feline side of the tale.

— . . . s-t-r-etch-ed o-u-t on his stomach . . . heAD RAISED . . . PERCHED [on the forecastle deck] . . . perceives feels and immerses himself in the SurROunDings . . . EARtuRN whistling from above [wind in the masts] the sound of water lapping from below [waves slamming against the hull] creaking sounds here and there [from the stationary boat], cries-flying overhead [birds] muted, abrupt shouting coming from all directions [humans] muffled chewing coming from below [rodents] . . . quIvEr-SnIffing the humid silt [the water of the port], new smells both strong and acrid [fresh cut wood, old paint, freshly applied tar] his well-known peers, the nibblers carpet agitated men . . . De-pupillate the cries-overhead, coming and going . . . "ex-tend claws" "invert claws" alarms, brusque sounds light swaying . . . FuR-Vibrations humid dry warm cold air . . . —

He hears, smells, and sees identical navy-blue figures circulating around him (officers in uniform the color of which cats would be able to discern) but whose eminences, clamor, and aromas—as much identical as varied—he can freely differentiate since they tend to be harmless. He shudders and shakes less and less as he notices some of them beginning to stop in front of him, getting bigger, growing in size, hum, then yelp, whine, bray, while licking at the aromas (odors and perhaps pheromones) that he senses as calming, reassuring him, inciting him to remain stationary and calm, even cultivating such a posture in order to repeat this sign of connection, which leads him to solicit their attention in other areas and moments on the ship.[7]

> You are of course aware that bodily postures are also an expression of emotional states. As such, they comprise signs addressed to peers in order to clearly express calm, fear, aggression, etc. Such signs are also sent to anyone else around, to other species who can also read them in their own way, understand or misunderstand them. In this case, humans anthropomorphize quite a lot, but such anthropomorphizing reinforces their belief, giving them more reasons for lending their attention to the animal. And they respond nevertheless, without wanting to, or in spite of their incomprehension, by way of postures, scents, and speech: by more signs. However, can behavioral modifications take place based off a pure misunderstanding? Has the cat not grasped, in his own way, that he elicits interest with his postures and therefore takes advantage of this, since humans establish a part of their favorable attitude based on these postures, and doesn't the cat repeat them in order to reinforce their connection, so as to maintain the benefit? Would postures then be not merely expressions and signs but also techniques of the body, chosen, adapted, and thus constructed, bio-psycho-sociologically staged, allowing for a positive interaction for the cat, and thus also for the humans since they take pleasure from it. You can see the interest in the vocabulary arising out of the human sciences. Whereas "posture" underscores an identical behavior for everyone everywhere (that is, what we used to refer to as instinctive behavior), "technique" presupposes a construction . . . within the biological framework but sublimated.[8]

"This animal cheerfully purrs," as Flinders defines and summarizes from the very second sentence of his journal, not frightened to approach him, to rub his legs against these sailors, integrating them into his world, making him a part of his own, constituting a community, which doesn't appear to be the case in regard to the other rat-hunting cats on the ship, all of whom apparently remain at a distance and independent. As such, Trim also simultaneously broadcasts friendly, sonic, and tactile sounds that the sailors notice (whereas they don't detect scents very well, even less the pheromones or do so unconsciously) and appreciate since they don't partake in the widespread cat prejudice of the era, at least as regards this feline whom they judge as quite different: filled with a bundle of "sweetness" and an "inner bounty," a "superior intellect" from his earliest years, with "extraordinary personal and mental qualities."

Don't mistake this for mere anthropomorphism. The sciences have now proven that in a number of species, including the cat, there is an individual variability of cognitive and psychological performances resulting from the endless interactions between their genetics and the environment. Such evidence should lead you to distance yourself from the old belief in identical individuals in each species. The variability of performances follows that of differing personalities. The other cats on this sailing vessel certainly have their own personalities as well, but the sailors hardly pay any attention to these other felines, if we are to believe Flinders's writings. Trim's personality calls out to them because he also expresses himself in the direction of these humans, in an adapted manner.

More than likely, he forged his personality based on his genetic, or perhaps epigenetic, disposition, which he then deploys thanks to good socialization among cats and which he cultivates during his interactions with humans. We must also take into consideration the qualifications mentioned by Flinders, which he notes much later on, as a description of the state resulting from this process of the development of aptitudes, whereby that which is acquired enriches the innate. One of these two dimensions alone is not enough. Contemporary cats demonstrate that an early socialization with humans can diminish reticence and avoidance on the part of the cat and increase interaction and play. Furthermore, the enclosure of an environment, such as an apartment or . . . a ship, can also foster such adaptation. However, Trim's siblings, who experience the very same living conditions, do not follow his path. The individual must also be attracted, desiring and willing for such an adaptation. It's as a singular cat that Trim takes interest in the human world, that he invests himself, participates, and grows along with the sailors within a hybrid community nourished through a shared territory, of gestural exchanges and expressions, entangled reactions, that he therefore forges, through action, into his state of being, his relations and his world, incarnating what anthropologists and philosophers are only now beginning to posit.[9]

Each time the marine chronometer is brought up onto the forecastle deck (a new instrument at the time, still a rare sight to see used on sailing vessels of the era) in order to verify the navigation or to determine the longitude through a pairing with lunar or solar observations, the officers all sense that

Trim wants an explanation. He sits in front of the thing, observes its deployment, tries to grab hold of *one of the attendant sailor's hands*, listens to the "tick-tack" clicking sound, turns and meows, moving toward the *officer when he observes the sky and cries out "stop." Trim wants "to learn the art of nautical astronomy," Flinders writes, veiled in a forced anthropomorphic humor, continuing to emphasize the cat's interest in the human world, even if his motivations are certainly different*.

> Let's (you and me) take another look at the scene of this anecdote from Trim's perspective.

—ear-turn as soon as he detects the chronometer's regular movement; head-turn, pupillation: human stirrings, magnification, location of the ringing; pushes his paws outward, quiver-sniff, examining carefully, listening: noise on the inside; throws out his paws to open it, nothing, continues to pursue the sound; circles the area [*as if he was pursuing a living animal, as Flinders will note, surmising that* perhaps Trim must assimilate the ticktock sound of the chronometer with that of a rodent's nibbling or the heartbeat of some prey animal he would seek to eat], comes back, meows at a human [being manipulative] to open it, tosses his paws toward another human [the officer] when he shouts; eyes him in order to solicit, obtain, nothing; and do it all again the next time the marine chronometer is brought out onto the deck . . . — In this way, Trim participates in a reciprocal interest and gaze, in the formation of a connection, albeit a connection built off different readings.

> And it's these possible divergences that make it difficult to go from such concrete interactions back toward cognition, namely: it's always risky to deduce an entire community of human and feline relations on the basis of the instauration of a single relationship. Such a deduction is true or false according to each case.

Trim is attracted to any activity on the deck, no doubt not during the initial voyage but at the port, and then during subsequent trips out at sea, in particular during the changing of masts and sails. Flinders describes Trim's mastered demeanor, but the cat had more than likely already observed and created associations previously, in particular in regard to the same and simultaneous shouting by members of the crew wearing navy blue (the officers), as well as the bounding humans (the topmen) climbing up and across the guy-wires. Trim becomes accustomed to c-a-s-t-i-n-g and MoVinG his paws along with them, raISING and PUsHING them just as quickly, FASTENING them at the same time on the same level, PLACING hindquarters and tail (*on the topmast*), eyeing them without following (*onto the yardarm*), certainly recognizing the danger doing so might pose to him, to let his paws FALL at the same time as these humans. Perhaps he assimilated such activity to playing with his siblings, a similar play comprising leaping and running on ropes and wood, and wanted to prolong such enjoyment with these other beings, which his peers did not do.

Slowly but surely, such activity led him to reinforce and express an "atypical level of courage and confidence" vis-à-vis humans, but it also helped to consolidate a connection with them, most notably through acquiring an immediate reward: "Since he always found a good friend ready to pick him up in his arms and pet and cuddle him once the work was done," *the sailors judging that Trim wanted to support them in their efforts and the officers feeling the cat imitated them while watching over them, and that in both cases he wanted to give them a hand in their work and protect them*. Trim demonstrates that tagging along becomes essential for him since "if he didn't understand the reason for the stirrings on the deck, he would meow and rub his legs and back against the sailors, often risking being accidentally kicked, until he garnered the attention he was seeking in order to satisfy himself." The same motivations no doubt, from play to some kind of reward, lead him to accompany the sailors in navy blue while they climb and descend from the gangway, between the ship deck and the quay, even moving faster than them.

If Trim allows himself to be noticed as a result of misunderstandings, simply seeking some kind of benefit for himself, the sailors nevertheless hold on to *the belief* (partly real and partly exaggerated based on Flinders's mood) of the cat's quasi-human gestures. Trim is tolerated and subsequently accepted and eventually assumes his exceptional behavior since he maintains his official and original function as a rat hunter and does not impose any sort of radical change in his status, *which the crew presumably doesn't want to happen either*, but merely acquires an additional condition. In effect, he shows himself to be attentive *to any change*, and *as soon as a barrel is moved, at the risk of being squashed, he quickly seeks to hunt any potential rats underneath, "the enemies of his king and country," writes Flinders, freely making use of the anthropomorphic metaphor.* "In the storeroom housing the flour he proves himself even more tireless in his quest; he frequently asks to be left alone in the dark for two or three days in a row so that nothing will get in the way between him and his duty."

So, Trim meows or scratches at the door in order to get in once he perceives prey and refuses *to heed the regular calls or to pay attention to the other comings and goings* of the crew, waiting and carefully observing, crouched, stationary, pupils-spread, quiver-sniff, ear-turn, whisker mustache fur vibrating. He is one of those cats who hunts in the dark, of which only a small minority still exist today, whereas most prefer dusk. Perhaps the rat-hunting cats at sea are more adept, *but Flinders goes to great lengths to emphasize Trim's status as a brilliant hunter.* Besides being well trained by his mother, Trim must have honed this skill through his audacious and tenacious character, a disposition shared by many contemporary cats. He honed this trait of patient waiting in his hunting of birds who would perch high up on the boat's ropes, or for the flying fish, who during *the "Reliance's" return to England by way of Cape Horn in 1800* would wash up onto the boat's deck and "for whom he had a particular taste."

His sustained aptitude for killing rats means that he can avoid being considered as useless or unsuitable since he attends to his job just like the other sailors; on the contrary, it will lead him to be promoted as "the favorite of everyone on board," from the officers to the crew, and eventually he will be transformed into the mascot of the ship as was common for a number of British ships between the nineteenth and twentieth centuries. He hears himself referred to as Trim more often and probably reacts better to the specific sound, *a name chosen by Flinders echoing the good servant in Laurence Sterne's novel. But we don't know the exact date when the name was bestowed upon him, no doubt when the lieutenant became one of his privileged interlocutors.* Since he finds himself well fed, outside of his own hunting, which he mostly continues out of pleasure for tracking and catching than out of a necessity for eating, he lives a different life than the other cats on the ship and thus acquires an exceptional stature for the era, *according to Matthew,* growing to a size analogous with that of a domestic Angora living in the most comfortable conditions, with a weight between ten to twelve pounds, still superior to that of contemporary cats.[10] He thus acquires a particular status, living like a house cat without truly being one.

He lives this privilege on a daily basis since he is allowed to partake *in the officers' meals.* He becomes accustomed to climbing up onto the table fifteen minutes before their arrival, no doubt following the emanation of the various aromas of food, luminosity, and his biological clock so as to know when the proper time has come, and thus imposes his presence. Is such a presence pleasing because he shows himself to be at once "sweet," that is, "well-behaved" in the eyes of the humans, as well as an "opportunist"? He learned to wait until everyone had been served before making his way around from one place to another in order to beg, by way of a "tender meow" that he knew how to adopt, *which they all wanted to hear. The officers understood it was useless to ignore him since* he pawed at their forks and the morsels of meat "with such dexterity and with the most gracious of demeanors that *he elicited admiration more so than anger,*" swallowing the food without fleeing or climbing down, sporting a temerity and confidence, *making them all think that he was merely retrieving his due. These men tolerated him and took pleasure in showing him off to guests in order to surprise them, thus creating a more relaxed atmosphere: for laughing, speaking, prepared to share the territory of the dinner table—one of the most intimate and human of territories—with the cat.*

This interaction, which became a ritual according to Flinders's depiction, presupposes some sort of adjustment, a reciprocal and progressive comprehension *not evoked by the officer but which he lets us glimpse through his anecdote.* While still a kitten, one day Trim sits on the lap of a human, who was simultaneously speaking and eating, stretched out his long torso, placing his paws at the edges of the mouth, and tore off a piece of half-eaten food, larger than him, no doubt repeating what he had done previously with his mother when she taught him how to eat prey. Early on, Trim only tried to sit on the lap of the human, not yet knowing how to articulate his request, *which wasn't well understood or received,* and, as he quickly fled once he had acquired the food, he finds himself punished in order to not do it again.

> You can perhaps surmise that Trim had to go through a series of norms and sanctions that were slowly formulated and imposed on him by humans. And as they were subsequently and slowly understood and assimilated by Trim, he was able to transition from the act of capturing to that of sharing, moving from conflict to that of policed connivance, accepted by every-

> one. This idea of norms, which implies the existence of refractory behaviors, is a major provision of the human sciences. It is the very condition and result of sociability among individuals from varying species, of symbolic interactions, and the construction of ways of being as well as individual cultures.[11]

This adjustment made by Trim also took place with others on the ship, such as the quartermaster in the dining room, whose "opinion of Trim's intelligence was such that he spoke to the cat as if it was his own child," such a rare occurrence that Flinders felt it necessary to write down. Such a proclivity presupposes that the human attributes a capacity and proximity to the animal in question, such that he extracts him from out of the grey depreciated mass of his peers. What happens is pretty much what takes place today, now having become widespread and studied. The interpellations from the quartermaster elicit Trim's attention, who carefully fixates on the sounds, on his gaze, on the posture of his face and body, on various odors, perhaps also on pheromones, *but Flinders mentions none of this, hardly noticing any of this at all as is common with humans.* Trim understands that they are speaking to him with favorable intentions and slowly adopts a posture, gaze, and a meow according to the circumstances, along with other signs *that humans do not detect. The quartermaster in the dining hall therefore feels invited to continue and accentuates his speech toward the cat.* Each of them adopts rites that seal and reinforce their understanding and relation: *the gift for one*, begged for by the other, of a ration of bread dipped in milk each morning; Trim's acceptance of climbing up on the shoulders, of rubbing his head against the quartermaster's cheek and head, as well as an exchange of scents, are all *signs of affection for the human*, but it is possible that Trim adds this additional dimension in order to get some morsels of meat.[12]

Trim and his sailors understand each other, adjust themselves, and collaborate even more during games and other learning opportunities that were still all very much rare for cats during this era, since it is in this manner that he acquires an "education far greater than what is normally granted to individuals of his tribe."— . . . instantly changes the direction of his pupils: toward something [a bullet] rolling around in stick-hands on the wood . . . l-a-u-n-ch-es paws extends claws distends claws loses: s-ta-rts-ag-ai-n moves clasps with claws closes claws, loses it again . . . Or ear-turn, changes pupil direction: things [bullets] arise in the hands from another human, rolling around on the ground, . . . launches his legs, gathering it up . . . Or perhaps moves his paws behind the thing . . . *too fast heads toward the human and takes off again* . . . turns head legs and paws around moving behind the thing again . . . *too fast heads toward the human and takes off again* . . . turns his head, body, legs again . . . —

Trim grasps and participates to the extent that he sees analogous emotions to that of hunting and feline games about which he gets just as excited and takes part in, or as *Flinders writes*: from which he takes "pleasure."

> A term that Flinders uses in order to anthropomorphize the animal, but that you can and that we must retain without worry on the condition that we make a contrary use out of it: we apply it to the specificity of cats. We must therefore note that this pleasure translates an accentuated state of feline well-being, physiologically and psychologically speaking, expressed by cats in the manner of cats: through a liberation of substances in the body, through scents and enveloping pheromones, through postures such as holding the tail straight, pointing the head and ears forward. As such, there is no reason whatsoever for reserving the term and notion of pleasure specifically for humans and to only concede the term of well-being to other animals. Short of wanting to hierarchize it by way of implying that there is more to human pleasure, whereas in reality, it's simply different. Rich and original certainly, but merely different. If humans don't express pleasure in the same way as other animals, they can nevertheless detect, deduce, and surmise the pleasure of other species they are close to, both biologically and culturally. They evoke it with their feeble human means, through anthropomorphizing, but this doesn't mean that they are inventing what they are detecting or poorly discerning it, that there is nothing to detect, or that they cannot detect it.

More than likely, Trim expresses his desire through turning, scratching, voluntarily meowing for the humans nearby; if not, such interactions would have never been developed, *but Flinders mentions nothing of this. Living in close quarters with these rat-hunting cats, the sailors on the "Reliance" have an abundant amount of time at their disposal while it's being prepared to observe how they act. And slowly but surely, both officers and crew take an interest in this one particular cat. They know how to enter into his world in order to propose games that attract him, whereas it would appear that they hardly bother with any of the other cats on board.*

> In the West we hardly played at all with animals in general, cats in particular, even kittens, deemed amusing to watch but hardly endearing, in the same way that we didn't banter with children. Such a situation seems very similar across the Netherlands, France, and Great Britain, even though the latter still seems to set the tone even today.[13] At that time, one had to be an eccentric, such as Christopher Smart, to entertain oneself and play with one's cat, whereas today it's highly recommended!

This is a good example for demonstrating to what extent Trim was extracted from his species and set apart, no doubt due to the unique, isolated, and withdrawn environment of a sailing vessel, where it was possible to remain in a contained and separate community [entre soi], so as to construct something considered as exceptional on land but normal on the high seas. Such transformation was also due to Trim's personality, which plays a role in the construction of games between him and the humans, for one shouldn't simply think such games were imposed on him as a result of a human order. In eliciting their interest, becoming responsive and then demanding their attention, he successfully places the sailors *at his service by having them dedicate their time, ideas, and emotions to him.* He constructs an oeuvre of a performing individual, active in the creation of his world, through his interactions and behaviors. He demonstrates this all the more through his participation in *human games* whereby he must enter into their world. Since "each man on the ship took pleasure in giving him instructions," Trim learns to leap over outstretched hands and "to lay on his back with all four legs spread out, raised in the air like a dead body; and he holds this posture until someone *signals to him to get up, while his preceptor follows in front or behind him.* And yet, if Trim held such a posture, which, we should emphasize, is not fun for a quadruped, a subtle movement of his tail indicated his tiring of this lesson, *and his friends never pushed their lesson any further.*"

> You will have noticed that these games are fairly similar to those practiced between Jeoffry and Christopher Smart or by you and your own cats. Such ludic interactions are deter-

> mined by the biological and cultural frameworks of the two species and through their zones of intersection, which suggest, impose, and limit the figures. These games rest on a reciprocal logic, of one toward the other, which has recently been highlighted by anthropologists and philosophers alike. They are constructed through the study of the world of the other, of his signs, gestures, and reactions. They are constructed through the deduction of possibilities for self and the other, by the creation of gestural interactions, circulating signs at multiple levels, from bodily gestures to sonic ones. When Trim signals his impatience by waving his tail, leading the sailor in question to respond by speaking, they both demonstrate that their interaction rests on a language that doesn't need to be identical (a fortiori human) on both sides but which is common, that is, understood by each one of them. All of this is obviously slowly constructed in order to arrive at a completed state, described by Flinders.[14]

With his/these capacities, Trim participates in a collective attachment that, in the beginning, doesn't appear to privilege Flinders. *The latter is present throughout the months of repairs made to the "Reliance" and during the ship's voyage in the spring of 1798, but his absence, while he embarks on the exploration of Australia on another vessel the same year of 1798, does not appear to have any negative impact on Trim. However, an inflection will appear during the other exploration, the summer of 1798, since Trim embarks on the vessel in question, in order to protect the bags of bread from rodents . . . at Flinders's request, as he writes with humor, providing a glimpse into a stronger relationship between the two, or with the brother of the lieutenant, present on the boat since 1795, who remained with the cat on the ship throughout the entire year of 1798 and who found himself completely immersed into the same interspecies experience but who is not mentioned once by Matthew. Moreover, it's thanks to a third man, the Australian Aboriginal, Bongaree—brought along to serve as a mediator but who had no other sailing responsibilities and therefore had a significant amount of free time—with whom Trim forms a strong bond* and an "intimate knowledge." Having certainly detected Bongaree's attention toward him, Trim adopts a nonhuman but common language, in order to make himself understood. If he is thirsty, he goes toward Bongaree, then hops up on a barrel of water; if Trim is hungry, he meows from the fuel depot and leads him to the stock rations. In exchange, Trim offers up more caresses, perceiving it all the better if they are appreciated by humans, which cats use as a method of appeasement and communion. Trim thus reproduces, adapts, and proposes what he had codeveloped with the quartermaster on the "Reliance," having found a human who is just as willing.

And we know nothing else regarding the cat throughout this voyage or throughout the return voyage of the "Reliance" to England during the first half of 1800. In October, when Flinders disembarks from the boat, henceforth made into a harbor vessel, he takes Trim with him, either because the connection had been solidified or because he had been declared the officer in charge of taking care of the ship's collective mascot. The fact that Flinders immediately hands Trim over to a lady at the port in order to take care of a number of other items—most notably getting married in 1801—indicates that it's more than likely the latter reason rather than the former, which doesn't preclude a progressive attachment.[15]

Trim finds himself a bit disoriented in relation to how humans and cats are used on terra firma. Having acquired the habit of observing from a distance and from high above (from the forecastle deck or the ship's maintop), he takes to climbing on the roof of the house, climbing through the open window and clinging to the wall. One day he even breaks through the glass of the window in order to leave the house. Hunting rodents without restraint, on another occasion he collides right with the china cabinet! And as a result, Trim persuades *the landlord—superstitious and suspicious of all cats, as was common at the time in the West—that he is the devil incarnate. Conversely, when Matthew takes him to London and leaves him to move about freely while keeping a careful eye on him since "there are delicate ladies susceptible to being frightened by such a strange cat," with his black fur and markings,* Trim elicits admiration, remaining stationary like a sphinx on the chair, hardly bothered by noise with much less swaying than he had endured aboard the ship and no doubt reassured by the presence of Flinders's scent at once singular and common with his own. *In London,* Trim is placed *with a friend, who quickly requests for Flinders to take him back since "he had never seen such a strange animal. I fear I will lose him.* Trim wanders out into the streets at mid-day rubbing up against the legs of passers-by." In other words, Trim does not act like a regular cat: suspicious when outside and preferring darkness as is the case with stray cats. Nor was he discreet and invisible, like other house cats.

He would have more than likely eventually adapted himself to the new environment, but he sets off again *in 1801 on the "Instigator," a ship un-*

der Flinders's command with a mission to return and once again explore the Australian coastline. Trim quickly finds himself "at home" again and rediscovers a good balance between his environment and culture. Throughout the long voyage between July and December, since the ship is old and takes in a lot of water—the British Navy retaining its good vessels for replenishing food and materials as well as war during this tumultuous period—Trim renews his ties with the ship's *sailors (an entirely new crew of eighty-three men, a dozen of whom are scientists or naturalist artists along with two lieutenants, one of whom is Flinders's brother). Trim benefits from his status as the commander's cat*, but he is also very active since "his sweet nature and confidence, tied together with his entertaining silliness, quickly make him the favorite among his new cast of comrades." Flinders does not provide us with much else. However, two other anecdotes serve as a window into the more concrete nature of the living situation.

Trim carries himself as the dominant figure on the sailing vessel, which includes several dogs, and makes it his own. When he wants to make his way through the group of dogs playing on the outside deck, he adopts "a majestic attitude," raISinG hEAD FUR TAIL LEGS in order to MAKE HimSELf BIGGER, A GiGANTIC MASS to INTIMIDATE, ScRATCH nose eyes of the one in front of him. *To such an extent that Flinders uses him several times to chase the dogs away from the forecastle deck where maneuvers are taking place: with his orders given to him, Trim swoops down from his promontory of surveillance (from the poop deck, the highest part of the ship, reserved for the officers)*, meowing and clawing, even leaping onto the back of any who are recalcitrant. "Trim continued his pursuit until the dog took refuge down below on part of the outside deck."

Is it this same sentiment of dominance that leads Trim to one day steal some lamb shoulder from the crew's dinner table with the help of another rat-hunting cat on board (this ordeal is the lone recorded instant where we glimpse the emergence of Trim's interactions with his peers) and doesn't try to flee once the quartermaster surprises them, grabs hold of Trim, and "brutally" beats him? In any case, such an event shows that the rapport he shared with the quartermaster on the "Reliance" is not the same as it is on his new ship, or at the very least, that Trim must unveil a series of approaches, negotiations, and adjustments in order to regain his status as ship mascot. *But Flinders doesn't tell us anything.*[16]

And the rest of his biographical narrative doesn't teach us much else about Trim's activities during the first and best-known maritime tour of Australia from December 1801 to June 1803. The "Investigator" at that time no longer being seaworthy, Trim embarks with Flinders on another ship to head back to England, making their way back by way of the South Pacific in a sailing vessel called the "Porpoise." The ship crashes into a series of unknown and unmapped reefs in the Coral Sea on the night of August 18, 1803, in the middle of winter. Amazingly, Trim survives the shipwreck, no doubt in large part due to his mastery of water. He spends two months on an island *with a part of the crew*, protecting their provisions from rodents, *while Flinders heads back out to sea seeking to find help*. Then, Trim follows him *onto another ship, the "Cumberland," which is headed toward England by way of the Indian Ocean, but this vessel begins to take on water and is forced to seek refuge on the Isle de France in 1803 where the English officers are arrested given that war has been declared against France. Fearing that the guards will eventually steal his cat, Flinders confides him to a lady in 1804*, but Trim will disappear two weeks later. For Matthew, it is probable that, given Trim's ease around other humans, he was probably not suspicious enough and was captured, beaten, and eaten *by a starving Black slave who had perhaps escaped and was living off any animals he could find and hunt, which was a common occurrence.*[17]

> Flinders's writing takes an exceptional interest in depicting Trim's environment, allowing us to think their relation, to glimpse the endless dynamic synergy between them. However, this narrative doesn't do much in the way of evoking the psychology of the cat or the other humans. And Flinders hardly ever writes about himself, evoking more the community at large rather than a specific individual. No doubt this is a result of the times, where there was a modesty about speaking about oneself, humor only helping to reinforce a mask, a modest approach where beings are indirectly suggested through the context and their actions. We have a better time perceiving the individual weight of existence in texts by other writers who in fact privilege it. Such is the case in the writing of Athénaïs Michelet, who evokes two transculturations: one regarding the conversion of Toto who, during a long-term stay of this human, goes from being a rat-hunting cat living in a boutique to that of a house cat and who will eventually change his status again after her departure; and the second regarding . . .

CHAPTER 5

Amplification

. . . older cats, already wholly formed, who do not entirely convert their behavior but who still increase the intensity of their culture of being a human companion.

Such cases raise the question of changes in behavior throughout a life, a possibility that was for a long time not considered possible for animals in general, and cats in particular. Ethologists have often believed in what they refer to as "programmed natures," where animals would examine their environments but were impermeable to them, varying neither at the level of species nor at the level of its categories, in this case: transforming from a house cat to an alley cat. Other ethologists noted the initial education of the young in order to adapt to their biological nature to a given environment, while simultaneously postulating that this initial education ceased early on and then the individuals would no longer change afterward. The notion of an endless construction, arising out of the human sciences, which the ethologists will also arrive at, in their own independent manner, will put an end to this old notion, allowing for a reconsideration of how animals change throughout their lifespan.[1] One shouldn't substitute this third dimension of adjustments for biological nature and any initial formations, but rather add it and take into account all the dynamic connections between the two. Such adaptations are best observed during a pronounced change in the environment. Such as that experienced by . . .

Minette: A More Familiar Time
(Neuilly, 1849–1851)

. . . who was adopted by Jules *Michelet in his Parisian apartment, near the Collège de France where he worked.* In reality, she was actually invited, having been left at an early age on the streets. For, according to Athénaïs, she had the typical allure of "alley cats who had not at all been crossbred": "standing too upright on its legs," the fur "short, strong, and resistant," with a "uniformly brown coat, with vague, wavy stripes across the hindquarters," svelte, with a "beautiful thin frame," muscular and "fibrous," making her light on her feet, agile, and supple, "the nervous life within her seems to have absorbed everything." She was quite different from Mouton, mentioned earlier on in this book, who nevertheless gave her a warm welcome.

After having received her initial education as an alley cat whose intensity we are not made aware of, Minette will thus undergo a first transformation in order to become a house cat. Minette's transformation into a house cat is evoked too abruptly in order to have truly been studied by the Michelets. However, *according to Athénaïs, relying on the writings of Jules*, Minette quickly adopted the nonchalant manners of Mouton. No doubt, she was drawn to and motivated by this friendly cat, given regular food rations and provided with calm living areas, all of this granting her a tranquil security, in contrast with her former calamitous and noisy daily grind of the alleys, which had forced her to be in a state of vigilance at all times. She demonstrated a clear preference for Michelet's "study, closed off, silent, far away from all the noise found on the street," partaking in "long naps throughout the day," resembling "lethargy," which were in fact experiences of deep sleep: signs of a profound calm that doesn't prevent cerebral activity, most notably dreaming. Minette loved to climb up and lay on top of the manuscripts glistening in the morning sun on top of the desk, not so much to ingest the mind of their author as Athénaïs liked to believe, but in order to enjoy the direct sunlight that reverberated across the white pages, thus warming up her body, which had cooled off through the inaction of slumber, especially given that her fur is not the insolating thick coat like that of Mouton. We don't know whether she was allowed to leave the apartment.

> You will have perhaps observed that contemporary cats sleep quite a lot, on average 66 to 75 percent of the time in a twenty-four-hour day, with 30 percent of that time being in deep sleep. However, this can vary depending on their personality and environment, with inactive cats left inside sleeping more often, which is probably the case here with Minette. In addition to the biological dimension, we also have the addition of a psychological dimension, both at the level of the individual and at the social level in relation with the environment—in particular with the other living beings. There is also a cultural dimension, since the individual constructs his or her practice of sleeping through synthesizing these dimensions and can also modify them. In other words, the individual use of sleep is more than likely a bio-psycho-socio-cultural activity for cats, obviously specific to the ways of felines.

Minette therefore does not spend much time with her humans, *Jules Michelet and his uncle Narcisse, a former typographer who Michelet had allowed to move in with him at the time of his retirement at the age of seventy in 1846. And considering Michelet's family pretty much left Minette alone, she didn't seek out their attention much either. Jules "never focused his attention" on an "education"* for the cat. *Athénaïs never makes mention of the eventual daily use of the cat's first name, like Flinders does in regard to Trim. However, the simplicity of the word "Minette," which indicates a connection without going too far, makes one think that the name was used to directly refer to her, notably at mealtimes, which must have constituted the essential of the relation. For these humans, as with a number of their contemporaries, this cat represented a presence, a family pet—a chat de compagnie—and not an animal companion,* and the same can be said about her relation to them, as she was certainly more attached to the territory with which she reconstructed herself, more than likely conflating them with it. Their case allows us to seek a more nuanced understanding of the scientific claim that inside cats interact more with their humans and are more attached to them than those who reside outside: if this is perhaps the case for contemporary cats in the West throughout the twentieth and early twenty-first century, it's far from the case 150 years earlier.[1]

One fine morning, Minette experiences a brutal rupture with her environment (the morning of March 12, 1849, on the wedding day of Jules and Athénaïs):—suddenly intrigued, she either DE-PuPillates, raISeS her HEAD poINTSEARS and WHiSkeR, or becomes worried SPreAds-DiLaTes, FALLs HEaD EaRs WHiskERs: cries, scents, hordes [of movers coming and going]; marked well-known objects moving disappearing; numerous unmarked empty spaces; suddenly senses being picked up and dropped into an enclosed space, grabbed back up, set on top of a marked objects [onto the final piece of furniture for the final voyage]; and all of a sudden: wobbles, braces herself with claws out; a hammering stench in front of her [*the galloping horses*], rolling thuds beneath her [*the wheels of the carriage*]; loud cries sounds enveloping odors, old familiar experiences ["*Industrial Paris*" with its "*honest and hard-working streets where everyone puts a bit of their life on display*"] then, encountering muted sounds and attenuated aromas with a frequent cavalcade of horses and people [*the western bourgeois part of Paris*]; and finally, arriving to be immersed into calm yet stressful exhalations and unfamiliar surroundings [*the gardens of the quartier des Ternes at Neuilly*].

Minette is free to roam in "a little house," where those she recognizes (her cat peer, Mouton, and her two humans) are all enveloped in unfamiliar forms, scents, and vibrations, whereas when an unknown person arrives, her great green eyes grow bigger than "*my marriage dress.*" Mouton—submerged as well in the new environment, unable to react or express himself (a common defensive reaction for cats under stress due to experiencing such brutal changes)—seeks refuge under the furniture for a while in order to grant himself the time needed to gather himself and evaluate the new shelter, spying, listening, and sniffing from a distance. Neither defensive nor aggressive, Minette expresses her stress through being "nervous and impatient." She leaps up on chairs, meowing in front of the armoires, "cautiously entering" into the study, "sniffing all the objects, seeking out a familiar and reassuring smells," confirming that they are indeed part of her world, making use of them in order to provide herself with location markers so as to carve out and master her new living area, that is, rubbing against them in order to reaffirm their connection and slowly smelling other objects and other places in the house and their saliant qualities to record them in memory, so as to situate and distinguish them, and to eventually rub up against them as well to assimilate them.

She must do all this within a new atmosphere, which requires her to adapt. She incessantly de-pupillates her eyes, reducing the size of her pupils in order to deal with the excessive brightness of the new landscape compared with the previous one, which no doubt makes her feel less secure than the former relatively dimly lit locale, *along one of the "narrowest and gloomy streets in Paris with no sky to be found," in a Paris before the arrival of Haussmann and his architectural designs and inside one of those bourgeois apartments with "thick wall-paper," providing a sense of withdrawal. Moreover, Athénaïs, just as annoyed by the bright light entering into the home, decides to fold back the shutters on the windows to the size normally reserved for spring in order to plant fast-growing plants as a façade, since "the outside whitewashed walls surrounding the home reflect the sunlight with an incredible intensity."* Later on, Minette will finally enjoy a reassuring midday *ambiance, since "the windows, through the curtain of morning glories and somber cobea flowers, will become, more than she realizes, luminous gaieties from the outside. The room was peaceful, calm, and cozy."* Initially, she ear-turns toward noises she had perhaps forgotten while living in the apartment: whistling sounds [of the window frames], the pitter-patter [of rain on the roof and windows]. Above all, she quiver-sniffs an atmosphere that is ceaselessly disturbed; whereas previously she lived shut off in an area with fixed and localized smells, she now had to deal an onslaught of stronger and stronger odors [springtime] that burst into her living space, invading, turning, and circulating throughout her environment and then just as quickly vanishing.

Once outside, Minette discovers another world. She takes to "vibrating," "often acting bizarre," and "even disagreeable":—observing with all her senses . . . drawn everywhere and by everything at the same time . . . by diverse forms, brusque movements, dispersed sounds, and whirling aromas . . . quiver-sniffing to one side, ear-turning toward another, pupillating elsewhere . . . launching her paws out here, becoming startled over there, freezing her stance, quickly turning her head in another direction, and setting off again . . . —Little by little, she identifies, locates, and maps out a territory, associating appearances, smells, sounds, climbing atop moving heights (across the branches of fruit trees) down onto varied paths, observing from above, spying on the flowing cries nearby (of birds "lively and happy"), shuffling between wide paths (of the bushes) or "the penetrating aromas" from below (flowers).[2]

And added to all this novelty, there is also the humans. Not in regard to the most slender one (Jules), mostly silent and still as before, always hanging out in the same area (his study) every day when the light is the brightest, then disappearing as the light vanishes; nor is there much change in the human at the lower level of the house (*Narcisse, who is barely even mentioned by Athénaïs, tiny of stature, discreet, and more than likely remaining in his bedroom in order not to disturb anyone, similar to Loti*). Rather, the novelty concerns the new human in the house. Minette can only connect her to the quirkiness of the new space since she perceives her walking for a long time outside, as she as well experiences an *upheaval in this change of environment, enjoying "the true delight of breathing in this new air" after years spent renting a room.* No doubt, Minette became accustomed to sensing her loitering around her, in the same way that *Athénaïs becomes accustomed "to seeing her* scampering around me in the garden, appearing quite busy, rummaging about the flowerbeds." While inside, Minette can observe this new element, coming and going, climbing the steps, descending the staircase in order to organize "*this little world that now belonged to me*," perceiving Athénaïs as being just as novel as the new environment. And yet, Minette hears her, smells her, nevertheless deciding that Athénaïs is similar to the others: still or rarely moving, quiet or meowing in her own unique way, upstairs, when the light is shining at its brightest: *reading or sewing while singing in her bedroom or Jules's study on the first floor in the afternoons.*

More than likely, Minette's interest in this new human is likely the result of her animating the environment. First, she pupillates Athénaïs from a distance—observation being the very foundation of learning for cats. She distinguishes Athénaïs's *face and records it to memory, studying her gestures and postures as well as the shape of her body, between a slow pace and moving quickly.* Minette more than likely also ear-turns toward her so as to assimilate the intonation of Athénaïs's steps and the register of the noise of her objects: the sounds of her voice, their modulation according to the adjustments made by Athénaïs, including the changing level of brightness and other presences. Minette will quiver-sniff her, intercept *her ephemeral* scents in order *to confirm her community with the others*, establish *her particularities*, take note of *her variations*, inquiring about *her state [of being].* Then Minette will begin to follow her from a distance, integrating herself into a game of reciprocal surveillance: "*All I had to do was sneak away a bit in one direction: and I was sure that if I turned around, she'd be just a couple of steps behind me. If I went back upstairs to my bedroom, it was quite rare* to not hear a soft scratch at the door: Minette had come to pay me a visit." She slowly continues to get closer to Athénaïs, first making "one hundred tours and detours" in the living area, in order to keep her distance while still being able to evaluate any reactions, then slowly getting closer to examine her, in particular during the morning routine:— . . . LEAPING up onto darker areas [armoires and closets], raising or letting her paws fall onto the coarse surface of the other [wooden shelves], FREEZING, SITTING, puPILLATING, CRAwLINg into holes [the drawers], quiver-sniffing the various elements being taken from the armoire or placed inside it . . . —

> I want to draw your attention to the fact that this rupture experienced by Minette illustrates the capacity of suggestion from an environment, namely, what is now referred to in psychology as the notion of affordance, a term now also used in ethology and the social sciences. There is a very good reason to carefully examine the events concerning Minette. The environment's radical novelty provides Minette with new possibilities for action, inviting her to partake in them as she slowly traverses, perceives, and records it, while also inciting her to engage in an entangled yet common history which codetermines and co-constitutes them. From now on, in this place, Minette is no longer the same, the same way the house and the garden are no longer the same with her presence. First, there is one particular element and actor in this shared environment—Athénaïs—who offers Minette the possibility of an original interspecies encounter. Athénaïs's attitude here indicates an attempt to truly engage with her (and the engagement on the part of Minette is just as important). While their perceptions, actions, and postures are quite different, they are close enough so as to be understood by each other: to encourage getting closer to each other and subsequently partaking in interactions.[3]

For, after having spent some time with Athénaïs, Minette begins to seek her out. She begins to jump up onto *Athénaïs's desk while she is writing*, initially making use of the height to see what is going on outside or elsewhere in the area but eventually turning her attention back to the human, who *lets her do as she please, not wanting to intervene*. She quickly adapts a "soft mrrrr sound" so as to signal her friendly state and to prudently get Athénaïs's attention, "gently offering a tap of the paw" toward the *dancing pen*, or perhaps in reaction to what she considers as a game she herself is proposing, or in order to elicit *a gaze, words, and gestures*.

Minette takes to adjusting her signifying bodily techniques for Athénaïs, an adjustment she must also expect in return from her new human friend, attitudes she interiorizes once she discerns their positive effects. Once again, we should recall that dispositions or suggestions alone are not enough to make her do this, since Mouton, who is less fearful and less curious, maintains a much more autonomous but confined life. One must also want to engage in such adaptive activities. No doubt, given the time period, gender entered into the equation. Female cats are considered to be more "gentle," more "friendly" than the wilder male cats, to the extent that the neutering of the latter, which tempers them, is still a marginal practice and the spaying of female cats, which can render them fearful, is not yet practiced. However, in the end, it's Minette's personality that leads to interspecies engagement. Minette is neither affable and independent like Mouton, nor tempered and sociable like Trim; rather, she is more like an alley cat—both curious and cautious, at once suspicious and an opportunist, in relation to the environment.

And it's for this reason that Minette takes her time—*Athénaïs estimated around nine months, so until the end of March 1849, from the time of her discovery before truly entering into "strict society." Throughout, Athénaïs wants to remain indifferent toward the cat, more than likely trying to control her gestures and actions, whereas, deep down inside, she is more than likely drawn to cats since her early childhood. She probably also betrays her feigned indifference through various postures, scents, sounds, perhaps even pheromones and vibrations*, which Minette can feel and decipher since she directs her intentions toward this human, which she does not do with *Jules and Narcisse*.

> You see, in light of the absence of an articulated language, we must take into account all the perceptual means at a cat's disposal: not simply that of eyesight, given that we still don't know everything about cats and the large possibility that there still may be other unknown perceptual means at their disposal. And it's certainly by these unknown means that cats, aided by their cognition, perceive others (be they cats, animals or humans); it's how they perceive personality, physique, and at the very least, emotional psychology, and the instantaneous state of the individual in front of them, through which they decide their possibility of action concerning this other or during their interactions with him or her. Sociolo-

> gists have demonstrated the same evolutionary changes in humans, beginning with bodily language and then the advent of articulated language, which even began to substitute for it. There are a variety of reasons to make use of their analyses, obviously with the caveat of de-anthropizing them and adapting them to each species.[4]

Minette succeeds *since Athénaïs begins to respond to her, that is, she looks at her and talks to her, referring to her by name while simultaneously using the polite French "vous" grammatical form, as she would do with other humans, henceforth making these positive signs agree with those beyond her control. Such interactions are precisely the same ones we see other women perform in regard to cats that we have discussed earlier on. However, Athénaïs's own history and personality also play an important role, since she decides to educate Minette even though she is over one year old whereas the prevailing wisdom at the time considered cats definitively formed by that age. In other words, Athénaïs participates in the still very rare belief that one could educate cats, and the even rarer notion that cats could change at any age through human contact: "Those who live in close proximity with cats can't believe in this immutability" of behaviors. With such convictions Athénaïs herself becomes the first environmental condition in so far as she seeks to inculcate a feline model onto the cat that she, of course, deems to be better: each society, each human, and each era having their own ideal or picture in their minds of what a normal, natural cat is like and which is more or less consciously proposed or imposed onto animals. Athénaïs in no way desires any of the most typical scenarios of cats during this era: the independent cat who willing scratches or the "thing" of the salon—opposites sharing the common distinction of no human intervention. She wants a feline that freely gives of herself and attaches to her, hence Minette's initial passivity, while still requiring much of the human who must be prepared to respond. And in their dialogue, the cat must express her personality, her ways of being, but also accept being controlled in order to "feel the good and bad through a gentle caress or a punishment," in order to correct her faults, reinforce her "good qualities" . . . that is, those considered as such by the human!*[5]

Minette thus feels compelled and encouraged to amplify her relations with Athénaïs and change status, going from a basic pet in close contact with her to that of a sidekick. During the winter of 1849–1850, she dissoci-

ates her territory from that of Mouton—who remains close to the fireplace and pantry—and recomposes it with the territory "where her owner lived," on the first floor, between her bedroom and workspace, thus more than likely incessantly rubbing/marking the most salient elements in order to concretize and affirm it. She also watches over her new territory: not falling into a profound slumber like Mouton, with her ear turned toward the slightest sound, *according to Athénaïs*, no doubt quiver-sniffing the slightest of odors, pupillating the slightest variation in luminosity. She willingly hops up onto the *lap of Athénaïs, who sits more often now than during other seasons*, and rests there for long periods of time, in order to recuperate the warmth she has lost due to the cold temperatures and being still, or in order to build and proclaim a connection when she rubs/marks cold/soft elements (Athénaïs's dress) or warm/strong elements (skin), licking them while she also licks her own fur ("her dress that she smooths out at the same time she smooths out *mine*"), possibly in order to taste it, to evaluate human pheromones, or to let herself or ask to be petted, that is, to imbibe the scents annexed in such a manner and perhaps tune the vibrations and pheromones so as to create a mutual satisfaction.

Minette integrates herself into more dynamic interactions, for example: playing with string or tossed balls of yarn, which both she and Athénaïs engage in, the latter tossing and the former fetching, thus training each of them in new behaviors, a common learning practice, as well as reciprocal understanding and forms of knowledge, *whereas their era only reserves such games to take place between kittens and children, and only continued if the kittens are taught early on*. First and foremost, Minette engages in dialogues, just as rare as they were during Trim's time: learning to modulate her meowing, becoming attentive to the sound of *Athénaïs's voice, who speaks to her often and obligatorily casts, through these variations, signs of her state of being and intentions*. And here as well, she acquires forms of knowledge and adopts certain behaviors, such as to come running when she hears a specific vocal intonation or to manifest a certain feeling.

A more specific example is when *Athénaïs sings and reaches a higher pitch "the note of a violon" ("la," 440 Hz) but also no doubt increasing the intensity to the point* it was no longer bearable for the cat with sensitive hearing. "She climbed up onto my lap, her eyes fixed only on my mouth. If I continued singing, she placed both her paws on my chest, and became limp as if having a nervous breakdown. Her voice rediscovered a screeching meowing that, for cats, can express a number of things. Her eyes, in spite of the bright light of her surroundings, remained dilated, as can also be seen when a cat is in a state of apprehension or suffering. If I carried on singing, she would climb even higher, and with a firm, quite human gesture, she would place her two paws directly on my mouth, she put a seal on my lips." She doesn't shy away like most other cats would do, hardly connected to humans, and instead tries to express her discontent and change the situation in order to stay, using all the bodily means at her disposal, even inventing and adopting various techniques. *One could posit the hypothesis that Athénaïs misreads Minette's response in a fairly anthropocentric way. She doesn't take into account the auditory specificity of cats, believing instead in a sentimental reaction on Minette's behalf, including the idea that Minnette herself responds by way of melancholy serenades in the guise of her pleas. One could therefore presume that each of them reads the situation in their own manner. But this would be to forget* that cats can sense the emotional state of a human and react to it; in this case Minette perhaps connects her own auditory suffering with the *psychological emotional suffering of the human*, and therefore endeavors to put an end to both of their miseries, which to the cat appear as interconnected. In this case, Minette will not be able to grasp the *human's penchant for pretending!*[6]

She will continue to get closer to Athénaïs due to the influence of two recent events: the death of Mouton, who, for a time, she still looks for, before transferring her complicity with Mouton to the human; the second event is the death of several male kittens who were fathered by Mouton prior to his own passing. Minette doesn't try to shy away, she doesn't isolate herself like many cats would do, but rather remains within the common territory. She meows and scratches in order to *leave the cottage when the cook relegates her to the area without any authorization, since that's where cats and other animals should be placed.* Sleeping, calmly, she waits "*at the back of the soft cradle which my solicitude has prepared for her*" *when Athénaïs goes to gather her up and stays "present" beside her*, each time licking her hand before taking a drink from the offering of lukewarm milk. "*At midday, we had three tiny cats, three precious darlings,*" *Athénaïs herself being pregnant and living in what would be considered during that era a rare proximity with the cats.*

It's perhaps a confusion of roles that leads Minette to vanish for forty-eight hours during the initial stages of providing her milk to the new kittens (the third day), to instead seek out prey to feed herself. *Confronted with the whining kittens, Athénaïs drops to her knees and decides to become their "wet nurse," force-feeding with a sponge drenched in milk, forcing them to eat since they no longer feel their mother's teats, covering them with a cotton blanket as well, holding them close to her for the sake of warmth.* Upon her return, Minette cLIMBS up the stairs right back to her spot ("the cradle," and just as soon offers her "heavy teats" then hoPs up onto the *lap of Athénaïs at the first request* and remains there while the kittens sleep. This sort of pairing, an entanglement of roles, continues on while the kittens play, which Minette observes from above *but which Athénaïs incites and, in part, directs, making shadow figures appear on the wall, even creating a fake mouse made out of sheets and cotton, and buying a wooden puppet.* In this way, Minette becomes engaged in a dynamic synergy with a human who initially became the primary element of her environment and then an equal actor, slowly but surely constructing a world with her.[7]

This dynamic becomes inverted, and an increasing distance takes place as a result of two successive events. First, there is the adoption of a hyperactive and loud spaniel who Minette senses as a rival since she eats into the time spent with her and eventually partakes in the same territory as well. Whereas she had obtained the authorization to spend the early mornings on the bed with the humans, licking a bit of coffee and napping a bit alongside them, and initially leaps, claws, and hisses to prevent the dog from also jumping up on the bed, she must eventually accept his presence one day, when *Athénaïs gives in to the dog's whimpers.* And more importantly, Minette also experiences a radical upheaval of her environment when Athénaïs's son dies several weeks after his birth on July 2, 1850, leading her to lapse into a major depression. Athénaïs will become detached from the world of the living, the unjust survivors, in particular, from her companion animals, which "belong to her like things."

The chaos begins with Athénaïs's disappearance, when Jules decides to take his wife to spend some time at Fontainebleau for a change of scenery. This leads to Minette looking for Athénaïs outside and inside, wherever her scent is strongest (the bedroom), and then, to eventually remain in the area which calls to mind their proximity, waiting and sleeping there and only leaving to feed herself or to continue checking for her presence in the surroundings. In this way, Minette displays that she has connected her ways of being to her human, having become the primary element of her world, more important than the rest of the environment in determining her actions, without nevertheless, detaching herself from this milieu and her territory, since she had never traveled with her elsewhere, having never been able to distinguish Athénaïs from other elements in the environment. Hence the magnitude of the disruption and, conversely, the excitement upon *Athénaïs's return.* "But in recognizing my voice, Minette hurled herself at me; I had her at my feet. Without giving me the chance to identify with her, she transforms my dress into a ladder and hops up into my arms, attempting in as many ways as possible to kiss me in her own way: licking my hands, turning around, and raising herself up to be bigger and make it easier for her to press against my face, not merely using her face, but her entire body, electrified with emotion." It still should be said that such a rare and punctual connection during this era is largely a result of *Athénaïs's personality and the fruit of her activities.*

And yet, Minette discovers nothing but an inertia when attempting to renew their synergetic relation: she follows her everywhere, carefully examining her, hops up onto her lap as soon as she catches a glimpse of Athénaïs's *gaze*, rubs her head against her hands in order to untangle them, tries to get her attention through pawing at her, but hardly receives anything of a *response either through gestures or speech*, often simply crashing *into a still and silent element, whose scents as well have more than likely changed.* Struggling throughout the autumn of 1850 through to the spring of 1851, Minette reorganizes her ways of being. If she maintains a presence, *to the extent that Athénaïs calls out to her, prepared to doing nothing upon her arrival*, and if Minette still regularly seeks her company, no doubt based on detecting Athénaïs's (supposed) mood, without receiving much of a response and without being able *to modify her mood, even though we now know that cats can serve a therapeutic function, and perhaps Athénaïs would have been even more worse off in her absence*, Minette appears to reconstruct her autonomy, most notably in spending more time outside. And the more time she spends outside, the less reliant on humans she becomes, relegating her to the background of her environmental panorama, in part dissociating their shared spaces, thus redeploying her territory, reconstructing less anthroposized and more territorial behaviors.

And during the nightfall in *the spring of 1851, it's within such a context of withdrawal, and in part a return to origins, that Minette finds herself caught in a trap in one of the neighbor's fenced-in backyards*—proof that she still doesn't have a good grasp of these spaces. —. . . . . . crAwLinG among green [plants]..quIver-SnIFF attracting path . . . enter . . . teETH Pull . . . drying clacking from behind [the trap] . . . TURNINGher-BODY FRIGHTENED HEAD EARS MUSTACHE TAIL FUR FLATTENED . . . SCRATCHING, EmiTs M-e-O-W, SCRATCH-ING, EmiTs M-e-O-W . . . a glimpse of light . . . POINTSEARS EAR-TURN Pupillates A human becoming larger and larger [the gardener who tracks various pests] . . . gathERED UP BRACES HERSELF CLAWS OUT RounDeD BACK fur STANDING ON END tail whipping back and forth . . . grating sound [chain from the well] . . . DEScend ... Pu-Pils widening [it's becoming darker and darker] FURVIBRATING [hu-mid] . . . SuDDENLy crash padding the water CASTING PAWS OUT CLAWING MEOWING . . . DEScending under, STIRRING, DE-SCENding STirr[8] . . . —

Minette experiences one of those untimely and often violent deaths that appear to have been frequently suffered by a great many cats in the nineteenth century when one attained the status of an individual life. Among the six out of the ten cats described by Athénaïs and whose death is mentioned in her writings, two die early as a result of illness, and four others have their lives taken not long after. Obviously, such a frequency seems to vary according to a given environment. Athénaïs's descriptions represent well what could be considered as the urban landscape where gardens and fenced-in yards populated the milieu and where cats were often considered as dangerous marauders, more than likely due to the chickens raised there. The rooftops and city streets are not safe for other reasons. In contrast, Athénaïs foresees a long life until old age for Bibiche, a six-year-old cat living inside a Geneva apartment . . . as long as she doesn't contract one of the brutal and deadly diseases often contracted by cats and which seem to infect any and all categories.

> During the portraits of Trim and Minette, you will recall that I defined transculturation as the transformation of cultures and ways of being in relation to their environment. In order to adopt this term, which originated in anthropology, we must set it free, that is, liberate it from its originary discipline and de-anthroposize it since doesn't just concern humans.[9] In this case, transculturation is defined as the way in which a group borrows traits from another group, shaping them for its own use and in so doing changing them and perhaps leading to improvements. Within the register of e(n)otholgy, this

> term must have a general meaning in order to correspond to several species, while also risking a more nuanced specificity with every application since a cat is not a dog, and a dog is not a horse . . . Furthermore, the definition must be adapted to the fact that the contacts are no longer simply between individuals or groups of the same species but also between individuals of varying species, most notably between animals and humans. We can easily transpose the human version of transculturation (to adapt, to acclimatize, to enrich, or to replace) when we are dealing with animals of the same species. However, we must transform the concept in the case of different species, since the biological barrier prevents species from acquiring and reshaping traits *stricto sensu.* The two cats mentioned above show us that it's more a question of adjustments, innovations, and construction of one species (the animal) placed in contact with the other (the human). This has an influence on the latter, who experiences change alongside the animal, allowing for a convergence by way of interactions. We can therefore define transculturation as a process of the transformation of individuals or groups through contact, through a process of borrowing or through exchanges if they are of the same species. And through parallel adaptations if they belong to two different species, With the changes taking place over a larger span of time, thus forming a historical dynamic that is reversible or not according to each case. For lived transculturations are rarely practices of abandonment and transfiguration, more so additions, mixtures, and hybridization [*métissage*].

We are often all too unaware of this individual dynamic under way, since it is masked by the perennity of specific territorial types even today, as a result of environmental supports. And yet, these categories have experienced a large number of variations over the twentieth and twenty-first centuries. In every country, alley cats are no longer a necessity due to use of poison traps and concrete construction impervious to rodents. As a result, alley cats have become the target of a growing campaign of eradication like that experienced previously by stray dogs during the nineteenth century thanks to refuges, dog pounds, and euthanasia. Such practices quickly led to a large reduction in their numbers when considering other variables such as the size of the cities, countries, and their chronology. And yet, among the postwar ruins and rubble after 1945, for a brief period, Germany once again saw their resurgence. In a similar contrast to an overall decrease of alley cats in cities, the number of stray cats among suburban neighborhoods began to increase during the second half of the twentieth century. Their increase was largely due to a reduction in violence during this period and such cats have subsequently remained marginal or hidden from view still today. They also benefit from more frequent contact with humans since their increase in numbers in Western countries has led to what we could refer to as the practice of a kind of "loose adoption." Such a rise in a specific cat population arose out of an extraordinary abandonment of house cats (by simply closing off entrances and tossing them out into the backyard, they are much easier to get rid of than dogs). As such, this type of cat saw a widespread diffusion during the twentieth century as the rise of feline reached its peak, most notably that of the common cat. The cat population continued to skyrocket from the second half of the twentieth century through the turn of the twenty-first century, including an increase in breed variety similar to what took place with canines between 1850 and 1950. This led to a number of new cat breeds all destined for human companionship and their satisfaction and tastes for a certain variety. The geographical and social diffusion of cats reaches its apogee with the recent trend and insertion of pet cats bred specifically for human companionship (sometimes arriving from the city accustomed to a specific way of life) and often kept apart from their cat peers living on farms.[10]

But the increase in cats is not the lone factor in their normalization by humans in this ongoing dynamic of transculturation. A variety of aspects accompany the increase and reinforce it. One such important factor is the social promotion of a certain large cohort of cat owners who had once been decried, namely: women, who were finally granted the status of "normal" in the interwar period![11] Other important factors are the utilitarian aspects of cats (for hunting rats, scientific experiments, furs, and their coats, etc.) that some people will attempt to resuscitate in the 1920s and 1930s, as they still consider them as natural. Such views will then disappear in the 1950s–1960s, which will amplify the partial conversion of the species to that of a companion species.[12] Another important factor is the prevention of a number of contagious, infectious, and parasitic feline illnesses that had initially turned the cat into a health danger when they were first discovered in the nineteenth century and beginning of the twentieth century, and which led to forbidding their residence in apartment buildings as was the case and respected throughout most of France and Germany. In the 1950s–1960s, this danger vanishes thanks to veterinarian medicine, which has by then learned how to deal with such dangers, leading to the abolishment of such prohibitions (1970 in France).[13]

For the most part, the diffusion takes place through the circulation of the image of the cat (as noble, independent, unruly) forged by artists and writers during the nineteenth century. Furthermore, their reciprocal affinity with this image of the cat becomes introduced in the first half of the twentieth century while already being widely shared by intellectual rebels. From Colette to Cocteau along with Foujita, a large portion of men and women of culture own and exhibit a cat. Their actions and proclamations are popularized through an educational system that has become democratized along with an increasing number of media outlets leading to their representation of the feline being taken as a natural portrait of the animal at the dawn of the era of mass media. And this will have an enormous effect on the social diffusion of cats between the 1960s through the 1990s. For example, in France, it was often those working in intellectual and artistic professions, along with social workers and civil servants—that is, those with most education and most sensitive to articulating themselves in a cultivated manner—who privilege the cat over the dog. If we should be mindful not to exaggerate this tendency, the rate of dog ownership being about equal to that of cats, such a tendency still demonstrates how the independent image of the cat resonates with groups who dream of independence in relation to economic and political powers. Writings and polls from the era indicate cat owners often describe themselves as liberal, left-leaning, and anti-establishment, while willingly labeling dog owners as being too loyal to the "lapdogs" of the system, to those maintaining the established order, that is, right-leaning. The cat with his "anarchist" (or at least anti-establishment) image is therefore used as a way to project one's identity and as a sign of left-leaning unity.[14]

However, beginning at the end of the twentieth century, newer reasons start to also figure into the animal's striking ascent: the cat doesn't appear to be tame, to be interested in us, want to go outside, or to be put on a leash and follow us. Starting in the last couple of decades of the twentieth century, the rate of cat ownership parallels the increase of urban density, along with a lack of space, and largely exceeds that of dog ownership in the case of big cities and housing for one to two people. The cat responds to the values of independence and individualism of an increasingly fragmented Western society eventually culminating in the fragmentation at the level of the family. The cat is finally considered to be more natural and truer than the dog, which has become too humanized, civilized, while the old domesticated/wild divide that was situated within the rural world becomes increasingly erased and replaced by that of the natural/artificial henceforth considered to be located at the outskirts of cities, whereby the search for nature expands into the latter.[15]

Here again, it's the notion of an independent—almost wild—cat that assures the animal's success. Being in the cat's natural reality, humans theorize their nonintervention and organize it, maintaining a distance, letting cats do as they may, which once again leads the animals to not consider humans as a primary element within their environment, thus fortifying an independent cultural territory and . . . confirming to the portrait that humans sketch.

The fact remains that the permanence of representations is shaped by a historical dynamic producing shifts, leading to . . .

PART 3

The Rise of Anthropozised Cultures (Middle of the Twentieth Century to the Twenty-First Century)

. . . as the lives of pet cats have become less and less autonomous and all the more artificial, like that of their owners! Following the dog, but with a lag time that has been recently overcome, cats are undergoing an evolution that mirrors that of humans: this includes the industrialization of their food as mice are now a rare occurrence in their habitat, the invention of indoor heating and a bathroom area, with cat owners even experiencing, in some cases, the desire to remove their cats' claws. There has also been a rise in the number of cats traveling, as well as contraception through sterilization, an increasing medicalization, funeral services, etc.[1]
All of this generates a significant increase in the role humans play within the environment of cats. And the number of cats has only continued to increase over the past several decades. To such an extent that the primary factor that drives their way of being and their cultures is becoming less their territory than that of the human. What was once punctual and a minority has now become common, and pet cats now co-construct themselves as companion animals. Obviously, such a shift was slow in arriving and one of the first steps in this direction could have been . . .

CHAPTER 6

Cats as Friends

. . . that is, felines evolving within a circumscribed environment, where humans no longer consider themselves as merely simple helpful presences, feeders, or owners, but rather as friends. They deploy specific behaviors and a specific environment for their cat, which they reciprocally consider as a family friend. A good example of this type of cat is . . .

Miton: Confined, Dependent

(Paris, 1930–1946)

. . . evoked by Marie Dormoy (1886–1974) in a short thirty-page pamphlet entitled *Le Chat Miton*, written just before the death of the animal in 1946 but published two years later in 1948, with a preface by Paul Léautaud. The work is more a brief portrait, detailing an ordinary day in the cat's life, rather than an extensive biography. It also contains black-and-white photos of Miton, photographed by the world-renowned animal photographer Ylla (1911–1955), no doubt sometime between the inauguration of her Parisian studio specializing in animal portraits and her eventual departure to the United States in 1941. The photos were possibly taken for their inclusion in another collection of cats photos, also with a preface by Paul Léautaud.

A luxurious volume with a small print run, and therefore expensive, this homage to Miton constitutes one of the first volumes dedicated to a single cat. It also illustrates the increasing publicity for cats during the twentieth century, a couple of decades behind that of dogs, as a similar volume was published in 1903 by the feminist and libertine journalist Caroline Rémy (known as Svérine) depicting a dog, Sac-à-tout, quite similar to the brief work by Marie. For this ongoing publicity is still the work of a very contained urban, intellectual, and bourgeois milieu, prior to the much more massive popularization during the second half of the twentieth century.[1]

The cat that Marie gushes over, confined and dependent, is certainly not something new—Bibiche, mentioned a bit earlier on in this book, perhaps also lived in a similar manner in Geneva around 1868—but such a relation is still rare at the time. It will continue to become more frequent after 1945 following the rise of the apartment environment as a result of the rural exodus, urbanization, the construction of large collective living areas, and a more demanding human context.

Miton, born (around the end of 1929, early 1930) to a black male alley cat and a Persian cat living in a *bourgeois apartment*, had fled twice to establish another territory (in the concierge's area) allowing him to hunt in other buildings and in the neighboring gardens. He has "jade eyes," a "Persian orange coat," actually perhaps more reddish—Photograph: a profile picture, with his tail held high, head turned, a bit of zebra stripes in contrast to the vertical stripes in the front, horizontal in the back, and rings around the tail, sporting a giant curved mustache—perhaps similar to the cat's father, who bore a certain "swagger."

He remains quite a long time (*3 months*) near his mother in order to learn without being rushed. *According to Marie,* he will have learned from his mother, among other things, to groom himself, to rest in the good spots, and to always go to the bathroom in the same place—one of those of sawdust litterboxes that would eventually outnumber chimney ash due to fireplaces being replaced by other household heating mechanisms and the growing requirements for better sanitation standards. He also has time to adjust to the others: the cats, by playing with his mother and sisters; to the humans; and to the *present concierge.* Whereas his sister shows herself to be "reserved," distant, more adapted to becoming independent of the apartment, *as many humans would like,* Miton is "cheerful, friendly, and curious about everything."

This pleases Marie, who wants a feline companion and who, not adhering to the dominant model of the independent cat who merely participates in a simple companionship, on the contrary judges the female cat to be "haughty." This description should demonstrate the role of feline individual personalities. In this case, human choice is not merely one of a projection of desire. It is also guided by an animal reality. Marie adopts the male in the spring of 1930 and comes to collect him with a hat box. Curious, he leAPs right inside—and just as quickly realizes he is trapped therein. WriGgLIng, meowing in vain; suddenly senses himself being picked up and moved; perceives his orientations vanish, unknown scents and noises enveloping him; feels a sudden stop, silence, and brightness; sPiLLs out onto the ground, GaLloPs to the window, grasps that he cannot get out, jumps up onto a hard and higher perch [a chair], smells milk and jumps back down to lap it up, perhaps getting closer to the prior living conditions; LeAps Up onto a higher and softer perch (a couch), s-t-r-e-t-c-h-e-s o-u-t upon the soft and voluptuous upholstery, perhaps recalling to mind the maternal, and falls asleep[2] . . . —

"From then on, he was mine," writes Marie Dormoy. The forty-four-year-old provides him with a unique environment for the era. Heralding from a wealthy bourgeois family, she lives an independent life and does not marry, preferring to take lovers instead, such as Paul Léautaud from 1933 to 1939. She loves to go out at night, drives her own car, is passionate about the arts, publishes writings on such topics, and at thirty-eight officially begins working first as a librarian for the fashion designer Jacques Doucet (1924–1929), then as a secretary for the art dealer Ambroise Volard (1930–1939), and finally as a curator at the Bibliothèque Sainte-Geneviève (1932–1956). With such a lifestyle that began by way of a small minority in the

nineteenth century, fortified by way of the wealthy urban bourgeoisie in the interwar period, and democratized beginning in the 1960s–1970s, which endows this case with a particularly prescient aspect, Marie proposes an isolated milieu, with a lone interlocutor, desiring a closer and stronger relationship than was typical of the time, thus allowing her to co-construct another type of cat: "Bourgeois in every sense of the term," perfidiously notes Léautaud, who himself played the homeless man among his vagabond felines.

Slowly but surely, he begins to discover and occupy all the available spaces, traversing them, smelling them, and regularly marking them. Curious and idle, he maintains his attention over a long period of time: "prospecting," "listening, spying, observing"—*Photo*: Laying down like the Sphinx with his head ears raised . . . Ear-turn Head turn Pupillating to one side . . . paws DOWN ready.

> You are correct in thinking that these photos were provoked by humans. For example, by opening the refrigerator in order to incite Miton to inspect it. Or in some cases the photos were arranged, such as when a bowl was placed on a bedside table claiming to be captured immediately following the cat's dinner. And yet, these photos still reveal the adaptation and adoption of feline postures, attitudes aroused by humans but in no way under their control, revealing to us how each of them was slowly constructed. These photographs nevertheless grant us a piece of reality and bestow a layer of depth onto the animal.

In spite of his vigilant observations and daily verifications behind objects or during other brief moments of withdrawal, Miton hardly ever has an opportunity to hunt: there are mentions of some rare baby birds and two lost rats to which *Marie nevertheless grants a large place in her tale in order to reinforce the idea and image of Miton as a natural cat, whereas the entire building where they live has been chemically treated to prevent rats from entering: a process that was in the midst of developing during that era.* He is not allowed to leave the house by himself and doesn't bother wandering in the hallways or courtyards *since Marie fears for losing him, the same feeling we glimpsed previously in regard to Bibiche. Such a protective sentiment was perhaps fortified during this time period as a protective reflex against the apparently common practice of the eradication of stray cats (intensified by various municipalities during the second half of the nineteenth century and the first half of the twentieth), a protective reflex that will only grow stronger as time goes by following the ever-increasing emotional importance of cats for humans.*

However, Miton does discover an outside (the balcony), learns to lay in its proximity (by the French window), and meows in order to gain access to its opening *when Marie is present*, goes outside, and sits there—Photograph: staNDING, PaWS in FRONT raised (on the railing), tAiL WhIPPinG, earsPOINTed, HEADBEnt, EAR-TURN, puPiLLaTinG QUiVer-SNIFFing below. He experiences no physical contact with other cats, or with other animals, and hardly with other humans *since Marie lives alone* but endures the impatience and emotional attraction of males since he was "liberated, via operation, from any of the hazards of love." *Léautaud euphemizes* he was neutered, more than likely at a very young age, *but Marie never mentions a word about it.*[3]

> This operation had already become commonplace for a long time with domestic animals. However, it was a rare occurrence for cats due to the pain it could cause and the risks associated with it: yet, it was practiced by specialists in castration, largely employed to castrate cattle in rural areas, who ripped away, crushed, or cut off their testicles without any pain medicine, or performed by veterinarians who also performed unanesthetized incisions with a scalpel to remove them—a practice that still bore a significant mortality risk due to possible infections that were still poorly known about or not well treated. In the interwar period, the rise of the use of anesthetics and more elaborate techniques will lead to the development and expansion of its practice. Due to its prohibitive expense, the practice of neutering prior to this period will be limited to affluent urbanites as was the case here. But we now know that Miton's unique case is a precursor to what will become a commonplace practice in the future.[4]

In reality, Miton does partake in several outdoor excursions, but always in the company of *Marie, who carefully watches over him and restricts his movements*. The first one, — . . . lengthy luminosity, an intense heat (the summer), feeling trapped collected and shut up [in a basket] and shaken; claws out, quiver-sniffing fresh scents [the hallways]; ear-turn, "VEntUres Out" among the bustling sounds "gorges himself" on the various acrid stenches enveloping him [*the automobile, which was still a rather rare event for a cat during this period*]; tossed about, VENTuring clinging; stop silence brightness [the forest of Fontainebleau] . . . —"His surprise in jumping out of the basket was so strong that he was breathless." Disturbed by contact with the spiky grass, he lEaPS up onto a softer higher well-known surface (a chair from home that was brought along) and stays there above the ground "for a good hour his mouth gaping wide, his tongue hanging out, breathing, completely shocked," invaded, discombobulated by the strange stream of acrid, subtle, strong, and delicate scents enveloping him. — . . . stretches out one paw, palpates the coarse ground, lets the other paws fall . . . ear-turn: strident cries from overhead [birds], sharp circulating whistling [the stirring tree branches], circular buzzing [the insects] . . . slowly pushes out his paws, and just as quickly recoils them, pushes them out again a bit farther . . . CrAwLinG through dense wavy greenery [thickets], casting his paws out onto rugged high places [tree trunks] . . . quiver-sniffing ear-turning pupillating palpitating . . . — He claws in vain so as not to be enclosed once again into the basket. Then, during the subsequent periods of light, he lets out "long meows" in order to get let back out once again to explore the landscape.

He becomes accustomed to long durations in the dark and other noisy, warm, and luminous escapades. *Such as when Marie brings him along in the car to accompany Vollard for the summer cure of a thermal bath*, where he bounces about. Miton is only allowed out to explore new areas (the hotel rooms) while perched on her, *in particular to spend time in the thermal springs, both of them cheering up the curists by their singular manners*, while Miton laps at the flowing water (faucets) while seated high above (*on the rails of the fountains*). He clearly learns how to remain free but calm within a loud and rumbling place, which slowly becomes another marked territory, since he remains huddled at the lowest level (beneath the chair), hardly moves at all, and doesn't eat when Marie moves the entire Doucet library with the car during the exodus of June 1940 and subsequently returns to Paris as soon as the armistice is declared.

Miton adapts to a third place, *since Marie sometimes hands him over to Léautaud for the summer (beginning in 1933) in order to go on vacation by herself*. And on each occasion, Miton feels constrained by the "imperative defense of his owner, overcome by a fear so palpable that he loses his bearings or inevitably encounters a misadventure": attached to a four-meter- long rope while outside, allowing him to come and go, smell and observe, climb up onto things, and then left indoors once night falls in *one of the writer's living areas on the first floor of the house*. He learns to socialize with his peers at the neighbor's, with dogs, as well as the French scholar René Guénon, displaying a "most accomplished animal sociability," thanks to his character and the help of castration.[5]

Nevertheless, he often spends most of his time in his primary and principle territory—Marie's apartment—where he adopts, very briefly for an unknown amount of time, regular ways of being, certainly due to the human and physically restrained environment, quickly mastered, inciting him to take on a routine more than likely maintained through his biological clock, itself determined by . . . the state of the environment, specifically in regard to luminosity and scent. Barely awake, he will take on the habit of meowing in order to go outside [onto the balcony]. Upon returning, he will clean himself and then begin to beg for food. *Marie offers him her "pâtés," no doubt confectioned with a mixture of cooked meat and vegetables, as was the common advice, in order to replace prey that had become impossible to hunt while still retaining a certain suspicion concerning any kind of raw meat, deemed to be an indication of wildness.* Afterwards he would digest, observe, and slumber—Photograph: seated, one paw in front outstretched posed above [against the closed door], scratching everything in focusing on the objective at hand. He can still eat since Marie has adjusted his feeding according to when the luminosity is at its strongest [between midday and 2 p.m.], after having ear-turned toward metallic sounds, quiver-sniffed the bubbling aromas, pupillated the coming-and-going *of the cook, who seems to arrive in the morning to prepare all the food for the rest of the day, but we aren't provided with much in the way of details, Marie framing her narrative in such a way as to impose the idea of a couple, even erasing any mention of her lovers or regular visitors.*

He then takes a nap ("for a couple of hours"), choosing the specific place according to his desires and the environmental conditions, showing that he perfectly masters his territory: for the freshest air and shade, he seeks out a nook on the floor during periods of warm weather—*Photograph*: partly spread out on his side on the wood floor, leaning against a plinth, head upright; resting above ground (on the couch and chair, both of which are offered to him) during bouts of colder weather, burying himself into the folds [of the couch cushions], or those of a fur blanket ["*specifically left there for him*"] or place in a closed-off protective area—*Photograph*: asleep in a tiny wicker basket, his head resting on the edge, the lid brushing up against his back. Awakened, he runs around or observes for a while, perched from above looking down, ear-turning into the position of the Sphinx, pupillates the outside through an opening—*Photograph*: on a table by a window, seated, his tail curled with his paws out in front. Here, he more than likely is sensing the arrival of another mealtime, with the luminosity diminishing and the sudden return of new scents, repeated noises, and makes his way into the midst of all the calamity and[6] . . .

I'd like to add a parenthesis here. Marie Dormoy's writings are a rare historical example regarding the daily schedule of cats confined to living indoors, which ethologists still don't have much knowledge about. In fact, for a long time, they refused to study such examples of felines, considering them too intertwined with nothing but anecdotes, preferring to study their peers in the laboratory—deemed unfettered by such dregs—in order to penetrate the inner workings of a natural feline. Without recognizing or wanting to see that such a so-called natural feline doesn't exist, and that the biological part of the animal is always composed and entangled with various parts of the social, cultural, and environmental, wherever the cats might find themselves. There has been a recent recognition that the cat in the lab is not the Cat as such—and even less the family cat—and that if the cat in the lab is more apt at revealing certain aspects of the cat, most notably neurological, he also hides others, as studies on pet cats demonstrate from research beginning in the 1990s–2000s. However, these scientific inquiries are predominantly based off questionnaire given to the cats' owners. And in this way, they grant more emphasis on the responses provided by the owners. All the more so considering they each indicate that they carefully watch over and view their cats in a good light. And if there was ever a need for more reasons to refer back to historical accounts regarding cats and their relationships with their owners, then this would do nothing but legitimize such practices. However, as with the other historical narratives, the writings of Marie Dormoy hardly provide anything of utility for ethologists, who are most of the time reliant on statistics: the categories and durations of behaviors, distance from humans and their interactions with them along with the variations, types, times, and frequencies of such interactions. But Marie's documentation does provide us with glimpses of these interactions such as a cat's interest in observing the outdoors, which has occupied such types of cats for a long time. Her documentation also provides qualificative aspects that are the preference of socioanthropologists, such as the staging

> and performing of various aspects of being, with perceptions, postures, gestures, or even various relations with Marie.[7]

. . . calls out—*Photograph*: standing up on his hindquarters, stretched out as tall as he can be alongside the refrigerator, he tries to snatch a piece of pâté Marie is preparing with his paws. After eating, he becomes more active, playing, at once spurred on by his genetics (dusk is the privileged hour for cats) and the environment, the time of day when he is most at ease and assured of the presence of his human. He GaLloPs after objects *(cushions or balls of crumpled up paper) that Marie appears to toss for him*, pounces on them with his paws, collecting and returning them to her allowing for the game to continue; hiding and creeping up on her while *she pretends to chase him, Marie adjusts to his maneuvers in the interest of redirecting his techniques for hunting*. In order to reinforce community, express affinity, he willingly rubs up against her and subsequently receives *"cuddles" in return*—Photograph: high up [on a desk], tail raised denoting a sign of appeasement and affection, he massages his neck, ear, and shoulder against Marie's nose and eyes, as she is seated, writing, and who responds in turn by the nodding of her face, adapting in her own way to feline manners. Unaware of the nocturnal activities of the felines living in the fields or on the streets, Miton will also eventually take to laying down while Marie sleeps, choosing his spot according to the situation: outside [on the balcony] when it's hot; "spreading out, lounging and taking up the entire bed" when it's cold; and "during the spring and autumn, resting on my feet, on my shoulders, chest, and even my head, according to the ambient temperature," in order to compensate for the heat lost during the nighttime hours.

More than likely, Miton learns through trial and error (but Marie doesn't mention any of this in her writings) to express himself in his own individual cat manner in order to be understood and get things. With his paws raised, he meows in front of a bright luminous opening (a French window) in order to go out, meowing all the more loudly if he must wait, expressing his impatience and his comprehension of a necessary insistence. Then, he "lEaPs up on my bed, furrows into my shoulder and chest, tickles my nose with his tale and mustache. Compelling me to get up and let him out," only reinforcing this bodily technique, which is slowly but surely deduced, learned, and adapted to accordingly by Miton. He then begins to meow when he begs for food, no doubt incited by the strong odors and sounds around Marie's dinner time, as well as by his own biological clock, which becomes synchronized with these repeated moments that pretty much take place every day at the same time. However, he also learns to adapt these bouts of meowing to other circumstances and intentions. And he also adopts a slightly different meow, "which he cries out, once *we have opened the door*" and "which must be a courteous 'thank you' or nice 'excuse me,'" Marie thinks, in anthropomorphizing her reading, as humans are wont to do. But she does not entirely misread Miton, rightfully noticing how he adjusts his use of language, no doubt incited to continue to make use of it by noticing it pleases his human who immediately opens the door for him. And nevertheless, she also proclaims that "she doesn't understand Miton's language very well," which she laments since, "conversely he understands mine quite well."

At a first glance, such a remark appears to refer back to observations made by ethologists: humans have the physiological capacity to hear these sounds that share the same frequency but bear different intensities; however, they have a hard time recognizing and distinguishing these sounds from one cat to another, due to the individual plasticity of the latter to create vocal sounds adapted to each specific individual and situation. *And if they don't pay careful attention, humans can even have difficulty distinguishing between sounds made by the same cat.* In contrast, the cat seems to locate the state and desires of a human through the use of several different forms of detection. *Nevertheless, Marie's account is quite remarkable in comparison with other previous accounts, from the very fact that she is carefully attentive to the cat's meows, for which she carefully notes their differences and the occasions when they are made. One could make the claim that her account is perhaps better only by virtue of the "random" nature of the selection* of Miton's *case, or that the silence on the part of previous commentaries on cats regarding their meows doesn't necessarily mean that the humans didn't notice them. It's better to reason in terms of a historical dynamic and put forth the idea of a growing interest on the part of humans the closer they became to cats. Marie's relation with Miton would therefore be illustrative of a certain stage, one degree within this ever-increasing relation that would eventually end with our current era, observed by ethologists, whereby humans easily decipher the signal-meows of their cats but not necessarily those of other felines (due to their difference). Such a situation would therefore not be an intemporal given but rather the result of a historical construction and thus of a temporal state.*

As such, Marie would deplore her inability to understand the cat, not because she didn't succeed, but because she wasn't able to attain an understanding adequate to her own liking, trying as she might to be attentive to her cat's expressions and therefore also more demanding of herself. *To the extent that such a state on the part of Marie allowed her to take better notice* of the efforts and adjustments on the part of the feline. And we shouldn't try to reason on behalf of the cat in an intemporal, ahistorical manner. Emerging out of this particular novel and more intense environment, Miton no doubt meows and modulates much more than his predecessor, or perhaps even his contemporaries. In other words, both Marie and Miton live in much closer proximity, making themselves more expressive and more vigilant in their communication, as well as more daring and consenting.[8]

Miton becomes accustomed to lEAping close to his human—Photograph: on the breakfast table, WeaVinG between coffee cup and coffee pot—so as to "lap up the bowl of milk, conscientiously licking, with her rough tongue, the mound of butter." If the cook left the room for a moment, Miton wouldn't hesitate to toss out his paws—Photograph: across a kitchen counter, he casts out one paw in order to push open the cover on a pot boiling on the fire—and "partakes in his part of the feast." "If the refrigerator is poorly closed, he knows, through a carefully acquired knowledge, how to jostle open the latch"—Photograph: raised up on his hind legs, his back left leg on a support structure, his other no doubt posed on one of the shelves, looking inside, the door halfway open—, having thus located the area, smelled various presences, and observed the various ways in which to open it. He does the same thing to learn how to drink—Photograph: his front legs in the large kitchen sink, his hind legs bent resting on the ledge, he laps a trickle of cold water—*but we don't know who has turned on the faucet*. He is just as quick to explore if *the cupboards and armoires are left open.*

Marie consents and admits to being guilty of having poorly shut the cupboards and armoire, letting Miton believe he was allowed to explore them. She has no intention of punishing him and simply decides to be more prudent in closing them so as to prevent such events but does nothing to hinder his activity. She gives in rather easily: waiting each morning, standing by the open window, no matter the season, to see if he wants to come in from the balcony; lets him lay down on her at night and even lets him sleep on her bed during winter "taking up the same amount of space as a grumpy old husband." So it is that she compensates for her guilt of maintaining a situation for this cat that doesn't appear natural to her, a sentiment also shared by Léautaud who reminds her of it to the point of even including it in the preface to her book. Marie is also an example of how to provide a liberal education for one's pet animals (or even . . . for children) along with the requirements for the owners (or for . . . the parents), which was slowly developed in the twentieth century, first in more wealthy urban households. To such an extent that Marie grants stronger, more intense and privileged position to this cat, raising their relation to a human level. Moreover, she names him Miton because "that was the name they gave me when I was, like him, a tiny blonde who didn't worry about anything." She thus oscillates between the animal and the animal-child, in both cases destined to provide her "consolation." "How many times, during the saddest hours, during times of illness, and sleepless nights, have I heard him wake up and come stand beside me, in the total silence created by solitude, his sweet purring seeming to tell me: "I'm with you and I'm happy."

For, in being in much closer contact with Marie, Miton senses her, carefully observes her, and acts accordingly. *As for Marie, in being more attentive to him, she observes his activities and writes them down.* He perceives the postures and emanations of an excited state when she impatiently waits for a belated visitor, more than likely a lover, and "he waits with me, his head raised and ears sticking out." He has probably also mastered waiting for his human to return home, associating a now familiar sound (the arrival of the elevator) with the arrival of odors and sounds of other humans, immediately recorded in his memory after the first encounter and quickly noticed again at a distance on subsequent occasions. Miton quickly arrives at the site of apparition (the door) and "is never wrong," serving as an alert of Marie's impending arrival home. Marie discerns another emotional connection. Since Miton also detects her exhalations and her state of being, he purrs and rubs against other humans who Marie appreciates, while nevertheless remaining "aloof, indifferent, even hostile to any advances or attempts of others to pet him." In this way, he shows that Marie matters to him. And it's within such a context that Marie notes an event that seems to be neither made up or anecdotal. When, during her return home after a long time away, Marie finds herself stopped by the German army and displays her "despair" to get back to Paris, Miton never moved once from his hiding spot (under her seat)— . . . raISes PaWs heAD EArs . . . clAwIng [at a pile of objects]

PuShes RAiSEs PaWS . . . rubhis headagainst hers and herneck . . . purring . . . —having perceived and sensed not the origin of the stress but the *state* in which Marie found herself.[9]

It is possible to explain some of Miton's behaviors as a result of being neutered, which tends to make males more affectionate and sweeter. Nevertheless, that's not enough since, even though the nineteenth century saw an increased demand for more obedient cats and thus a subsequent increase in neutering—supposed to make them lazier, contemplative, and more inclined to obey—these attributes are not common to Miton, therefore implying such behavior was not an effect of neutering but one of environment. The *context of spending time in close contact with the human is much more important* for orienting Miton toward more interactions, even if being neutered probably helped to facilitate this. The individual character of this spry, sociable, and playful feline is just as important.[10]

Miton also appears to terribly miss Marie during her numerous absences working elsewhere outside the home, regularly preparing exhibits, publishing a large number of works, and participating in various other engagements, including vacations. *Whereas Marie hardly ever mentions this situation, preferring to evoke their close relation, thus minimizing this contrasting reality*, Miton most often finds himself alone. He appears to compensate by sleeping a lot during the day, as is the case with many cats in such limited, known, static environments, which do not stimulate them. He meows loudly with a roar upon her return, patiently waiting up until the wee hours, no doubt perceiving their imminence through the ever-dimming luminosity of the day or through the arrival of human scent and signals from his internal biological clock. And yet, Miton adopts other attitudes that could be the result and signs of stress.

> Those among you who house pets certainly know that the discovery and acceptance of the stress of animals is a major event of our era. Such an event has led to a large number of studies about which I will provide two remarks within a larger historical perspective. First, the discernment of stress is often associated with the idea of an abnormal situation experienced by the animal. This immediately leads to the question of choosing a specific normality. For the cat, would normality be that of their wild ancestors—but domestication took place somewhere in-between—or that of streets cats and cats of the fields considered to be free? Depending on the choice

one makes, the gap between normal and abnormal is not the same and the cat's envisioned poor well-being thus also varies. Furthermore, this search for a normality often refers to a stable behavior that would define the species, whereas history would suggest that behavior is the fruit of the adaptation of the biological framework according to changing environments. And this research leads ethologists and veterinarians to often forget that they are only observing contemporary Western cats. Cats from non-Western environments would not necessarily behave in the same way, and nor would their European ancestors.

As such, the 1995 publication of *An Ethogram for the Behavioral Studies of the Domestic Cat* by the UK Cat Behavior Working Group or the creation of an English internet site in order to allow for the emergence of a consensus regarding cat behavior are in part nothing but illusions. The behaviors defined are only those of the contemporary cat. Historical and geographical variations are not taken into consideration enough. Here again, there is not enough differentiation between what behavior is arising in a particular species, or a specific group (of a specific place or era), or in an individual, even if this differentiation is difficult to establish given how much everything overlaps.

Obviously, these studies allow us to better understand and palliate current issues around animal stress. To the extent that the variations of substances in the organisms concerned, secreted in response to malaise, prove that such problem regarding stress does in fact exist. The relativity of things shouldn't lead us to sardonic relativism. However—and here is my second remark—malaise can also be the result of the inability of the biological framework to adapt to the environment but also is perhaps the result of another difficulty, namely a delay in adaptation. One mustn't neglect the dynamics of adjusting to changing environments and simply decide what is normal or abnormal through a simple glance at the current situation, which produces neither a normal or abnormal cat. As a result of epigenetic adaptations, even genetic adaptations through natural selection, cats could potentially be able to eventually adapt to current situations that produce stress. And the present situation, with its particular stressors, would then only be a temporary situation. One must think in terms of a history of malaise and well-being.

> Each environment produces its own particular cases of each and the current examples are mere deviations in relation to a temporary norm, of historical states.

In the early morning hours, at the first signs of daylight, when he goes outside (onto the balcony), he slowly chews on leaves (couch grass)—Photo: his hind legs on the ground, his front legs in the flowerpot, tail upright, with a rounded back, head plunged into the twigs. *Marie analyzes his actions as* a way to purge himself, which is possible, but we now know that the repeated chewing of plants in confined spaces is a sign of stress, here a question of apprehension of Marie's disappearance or of solitude, if we conjugate such an action with others such as self-cleaning to which Miton dedicates quite a bit of time or anorexia. He can go *several days without eating when his human is gone for an extended time, leaving a neighbor to watch over him*. When he finds himself confined somewhere else (*in Léautaud's apartment*) during the warm season, he regularly urinates at a height (*on the chimney*) instead of in his typical area even though that area is still present, *which Léautaud rightfully analyzes* as a manifestation of "discontent," in this case, at being without his human.

He appears to experience and betray a situation of dependence on Maria, which was previously rare for cats, but which will subsequently increase culturally, including the apparition of separation anxiety, which has also now become a widespread phenomenon. *In effect, Marie doesn't want a simple obedient cat, something to keep in her living room. She incites a much stronger relation by imposing conditions adapted to this, which she is completely aware of since she concludes her opuscule by summarizing* this feline life "that I keep imprisoned, that I keep separate from my peers, that I deprive of my greatest joys and who, nevertheless, grants me with the grace of his love." *But she abandons him most of the time!* Miton lives a paradoxical situation of being incited to get near to this human, to such an extent that his dependence on her proves that his being, his manners, and his culture are determined as much by her as by the territory, while nevertheless living most of his time without her. And yet, he doesn't benefit from the wealth of the open milieu that would promote a more active life on his part, nor does he receive any medical treatment for his stress, which has only recently been developed.

And yet, Miton nevertheless will live a long life (sixteen years), a rare occurrence at that time, but which also becomes more and more common. Miton benefited from a good quality of life and living conditions, from a less stressful state than the constant search for food and the daily survival on the streets or in the fields, sheltered from other contagions, even if he experiences another form of stress. Whereas the amalgam of cats living with Léautaud die like flies, Miton also benefits from increasing advances in veterinarian care such as vaccines, which were reserved for the wealthy urban cat population like him. Upon his death, Léautaud will write the preface for Marie's book on Miton, but doesn't specify the cause or how he died on March 14, 1946. Marie will bury him in one of her friend's parks. She will experience a terrible sorrow, in line with their reciprocal dependence, a sentiment still present two years later inciting her to publish her book as an homage to him.[11]

> Marie thus experiences a long period of mourning, a sentiment whose expression is still rare in the West but which will become increasingly normalized. Her behavior is close to that of Léautaud, who populates his journal with obituaries as well as memories from birthday celebrations, even if he seems to get over him quicker thanks to his other animals. Marie's sadness appears to be more intense or, at the least, more expressed than some of her other emotions mentioned above, which were less discussed and more intimate or less felt and more restrained, or perhaps a mixture of both, and apparently exacerbated by the typically short lives and brutal deaths of cats incessantly replaced, which Marie herself does not do, remaining without a new cat, no cat being capable in the short-term of replacing Miton. And here, there appears to be a historical dynamic at work between humans and cats, forging the connections between them.
>
> Such a connection can be detected through a variety of sources, but its precise history remains to be mapped out. It can be glimpsed within the portraits and stories recounted in this present book, which note the new status granted to the cats mentioned. At the scale of the species, such a status should be considered alongside others such as that of obedience and companionship, two traits often referenced in regard to the human-cat relations in the West still today. At the individual scale, cats begin to replace humans: cats become friends, as is the case here with Miton; then they begin to be considered as children of the family starting in the second half of the twentieth century. These . . .

CHAPTER 7

Cats as Family

. . . spur more attention and more contact in both senses of the term.

To reflect on these new cats, we have at our disposal studies that have been under way for several decades by ethologists, veterinarians, and even socioanthropologists. However, the questions posed by these researchers are often very precise and are less concerned with cats at the individual level and more focused on the group level. So as to maintain a continuity with the previous individual cases discussed in this present work, it's perhaps better to focus on spontaneous, more immediate, and freer expressions from a variety of cat owners, on the condition that such commentaries are sufficient, interesting, and focused on the animal side or viewpoint, which is not often the case. However, many of these inquiries are restrained or fragmentary, most notably in the media and on the internet, which justifies sticking to accounts written down in book form.
To the extent that they take into account a specific reality, these latter accounts will experience a geographical shift to the Anglophone world with which we must compose our reflections. Whereas France once produced a significant number of documented narratives concerning the lives of individual cats, such a specific literary output greatly diminished over the last decades of the twentieth century. Professional writers suddenly stopped writing animal biographies, perhaps due to their social normalization or because they preferred to write more literary and fictitious tales due to

their surplus value or to make a name for themselves. And those French writers who do recount tales of cats remain loyal to the independent conception of the cat first depicted by the Romantics.[1] Such as the common masters who begin to publish such tales in the 1970–1980s, because they belong to the cultivated middle class and circulate the same literary references as the Romantics. However, even though such romantic portrayals of independent cats still reign in the French imaginary, this doesn't mean that the evolution of cats considered or treated as children or family don't also exist in France. Rather, this type of cat relation has only become valued above all thanks to the intermediary of short videos posted on the Internet by way of a third wave of cat owners who no longer bear the cultural heritage of the nineteenth century and who no longer compose interesting written accounts of their cats, at least for the moment.

The opposite is true in the Anglo-Saxo countries. Paradoxically, whereas Great Britain invented the genre of animal biography, with France following shortly after, it has hardly produced any marketable documents, and the same can be said for the other countries as well, writers preferring fictional accounts or those where the human is the principal focus.[2] These authors have hardly changed their style and any novelty in the type of accounts of cats has arisen from ordinary owners who began publishing in the last decades of the twentieth century.[3] Often unfettered by European cultural references, not believing in the stories recounted by the Romantics, they practice and evoke new felines. Such a new writing is all the more easily understood given that these new types of cats have also increased, for the moment, in these same countries. Certain of these new writers have provided a wealth of documents, such as the books discussing . . .

Dewey: Or Human Needs
(Spencer, 1988–2006)

. . . whose story is presented by Vicki Myron, a librarian and director of the municipal library of Spencer, Iowa, in the United States.[1] The sociological ties with Marie Dormoy stop there, since Vicki, heralding from a family of farmers, marries young, becomes a mother, divorces, returns to her studies (which she had put off as a younger woman due to familial obligations), works in an agricultural region dealing with an economic crisis, and resides in a milieu that is very different than that of the Parisian. Furthermore, her account is also organized and revised by a professional coauthor, Bret Witter, who is a specialist in these sorts of documents.
Let's pause for a moment regarding this sociological shift concerning those who are now documenting the lives of their cats: ordinary cat owners whose writings largely outnumber those of the professional writers. Many of these new amateur writers belong to the (lower) middle class, precisely the class that will normalize having a cat as a pet throughout the second half of the twentieth century. This sociological shift will not, however, stunt the continuity and ongoing development of cats as a companion species, which began in Europe; nor will it diminish the quality of written accounts about them, as the new class of amateur writers have maintained a desire to describe and carefully observe their cats.
Conversely, such writing also bears an inflection in the way in which the documentation is expressed, granting a larger place to emotions and autobiographical accounts. These two characteristics are cultivated in contemporary Anglophone countries—which was not the case for a cat like Flinders—by both men and women, all the more accentuated through editorial and marketing strategies so as to attract a larger readership that is sensitive to these specific aspects. In other words, if these new writers no longer disseminate literary references, something Marie still chose to do in her writing while doing the opposite in her actions, they nevertheless respond to another norm we must be careful to make note of. And this why the additional cat books published by Vicki Myron, after the success of her first book, will not be discussed. Given that such additional works were written merely to prolong the potential profit through an increasing readership, we fear that this new norm of profit-seeking overrides the initial spontaneity of the first book. As far as the participation of a professional coauthor, his interference appears limited by the fact that Dewey is a singular cat who spends his time in a library and is therefore difficult to place into the mold of the familial cat archetype even if he falls within the category of a cat-child. If we must remain prudent in our reading of a given scene, attitude, or the language used, it doesn't prevent us from grasping a singularity, specific manners, or a culture.

— . . . shh-aakk-inng . . . COLD FALling on him . . . cold spreading OVER HIM . . . Shrinking legs paws tucked in, head lowered . . . WeAving between things, smelling something of the human and the tree . . . tired . . . with a hollow belly . . . suddenly ear-turn noises human voices . . . m-e-o-w ["It doesn't sound like an animal. It sounded more like an old man struggling to clear his throat"] . . . an abrupt clearing DE-Pupillates raising HEAD . . . detects human scent that is not a threat . . . LOWErs head . . . exhausted . . . feels himself being touched, then picked up surrendering . . . —

Let's recount the entire event from the human perspective.

Upon arriving to work around 7:45, Monday morning of January 18, 1988, Vicki Myron is quickly welcomed by a "low rumble" coming from the library's metal drop box where readers may return their books to the library after hours. In opening the lid to the drop box, the first thing she felt was a blast of freezing air (it was 5 degrees Fahrenheit) that had penetrated the chute due to someone lodging a book into the opening; then she sees a kitten hiding among the pile of books. Melting with emotion, she picks it up and just as quickly notices his thin and weak body is shaking. *Vicki assumes he was tossed in the drop box as a joke or to save him from freezing to death. It doesn't cross Vicki's mind that the cat might* have leapt up and slid into the drop box by himself in order to seek shelter, even though he will subsequently demonstrate his ability to SLinK and WeAVe everywhere.

The cat will also quickly sense the warmth he can capture in order to increase his own—the scented warmth of a human [hands] . . . a mixture of soft and rough surfaces intermingled with perfume [human clothes and arms] . . . nestling up against it . . . a moist yet coarse sensation overhead [a bath towel on the woman's shoulders] . . . surrendering himself . . . shuddering due to contact with a warm liquid [a bath in the sink] slowly relaxing his warmed and massaged skin and muscles . . . purr . . . a strong sensation of dry heat blowing between his fur [a hairdryer] . . . once again a sensation of softness [arms] . . . —Placed gently onto a slippery surface [a table], he tries to get up, carefully observing, sensing, and smelling the (four) humans surrounding him. He detects a warm welcome in hearing their voices, vibrations, their pheromones and odors, since he carefully moves closer to each of them, rubbing his head (in order to mix with their scents) against their hands placed out as a friendly sign in front of them and purring as a manifestation of well-being and appeasement. Next, he immediately melts in the warm embrace (arms) of each of them as he is passed from one human to the next, having no desire to bite, claw, or flee, and— quite the contrary—gazes at each of them with a manifest interest.[2]

The employees take turns carrying him around for the rest of the day, and it's quite simply his general attitude, postures, his rubbing and purring, his gaze—read or interpreted as friendly signs broadcast to everyone—that lead Vicki and her coworkers to decide to keep him as the library's cat. All the workers agree to chip in so he won't be a financial burden to the library, bringing him a box to sleep in, blankets, and a litter box, naming him Dewey after the book classification system symbolizing his adoption and becoming attached to him. The first few days after his arrival, the workers are very attentive and sensitive to the cat's demeanor. And even though he's not yet familiar with the new environment, he remains confident, patient, peaceful, and sociable *largely due to the workers creating the very conditions necessary to keep him*. Here we are introduced to a fundamental role of animals, just as important as that of humans, in the co-creation of relations, and more specifically the effect of their temperament, in this case audacious, confident, and easygoing, which Dewey embodies in his own individual way.

Such a demeanor can make one wonder as to Dewey's origins. Is he a simple alley cat, *as the veterinarian they eventually consult believes to be the case*, based off his common orange tabby coat? But a cat with such an easygoing demeanor implies he had already experienced good relations with humans, much different than cats of the past, in such a way as to allow for punctual contact through which to forge such an adapted temperament. Such relations have certainly developed throughout the West, through the general increase in pet cats, and this also seems to be the case in Spencer, Iowa, where the arrival of a cat with Dewey's demeanor is in no way received as surprising. The veterinarian's hypothesis is therefore certainly plausible. Or is he a cat who is lost or who has been abandoned, therefore already having experienced interactions with humans, as his meowing inside the book drop box would tend to indicate along with this easygoing demeanor around them. This hypothesis is just as likely. *Whatever the case may be, such a relation between the cat and humans implies that the humans are prepared to welcome him. Vicki and the rest of the employees are quick to embrace such an alley cat. His situation immediately elicits emotions on their part and they just as quickly shower him with affection, no doubt ballasted by a general secular promotion of all cats. This case thus clearly demonstrates a completely different human environment in regard to cats than in previous eras. And if Dewey is indeed an alley cat, it also demonstrates just how different and singular he is as well compared to his peers from previous eras.*

For he makes himself just as sociable with his initial public, specifically the *members of the library administrative council who bear the lone responsibility of authorizing his presence.* He lets himself be cradled like a baby, whereas most cats do not like this inverted, agonizing position. He melts into their arms, rubbing his head against them, staring at them, and therefore giving off signs of solicitation and complicity in knowing how to accept and make use of such human ways. His attitude *persuades the council, as well as the local readers and press, to baptize him as the library's mascot. Insofar as, according to Vicki, the population of this agricultural region, ignored by the rest of the United States, would more than likely identify* with this abandoned alley cat.[3]

It will take some time for Dewey to recover and regain his strength (ten days) before he begins to explore this new environment: a vast enclosed site (1,200 m^2), partitioned into different spaces based on their form, luminosity, sounds, odors, humans, and objects, even though elevated surfaces abound (tables, chairs, and bookshelves) where he can climb. There are nooks where he can hide (among the rows and shelves of books). There are places where he can sneak around (closets, drawers, card catalogues). The entirety of the space oscillates between drab (concrete) and luminous (large bay windows), hot and cold. It is also quite chilly at night (when the radiators are shut off). But Dewey has attained an age (around eight or nine weeks) where he can self-regulate his body temperature, which no doubt saved his life when he was stuck in the library drop box. He is now able to master—like an adult—his senses, orientation, and gestures to the extent that either his abandonment by his mother or even an owner led him to quickly learn how to take care of himself. However, he is betrayed by his slender frame (Vicki thinks he is only about two weeks old), indicating that he poorly compensated for his abrupt and premature weaning from his mother and no doubt his weak initiation in hunting—a practice that he will never truly develop later on. It is also worth mentioning that the concrete structure of the library prevents rodents from entering and therefore will not serve as a good training ground for acquiring such hunting prowess, not even in the form of simulation as was the case with Miton.

Within such an environment, which serves as the determining factor for his actions, he would be able to become a simple territorial cat, peaceful but distant with humans. For he takes possession of this space wherein he can roam where he pleases and he constructs his territory with these spaces. He develops his daily pastimes, his favorite moments, taking advantage of the bodily plasticity of felines, developing it— . . . slowly casting his paws out deeper into the lair of different boxes (of card catalogues and Kleenex) . . . spending his first winter in the library mostly curled up in a box [full of free tax forms for the general public] henceforth replete with his scent . . . slowly slinking paw after paw into various holes [drawers] venturing up into protective areas overhead [adjacent shelves full of books], preferring to settle down on the highest rung of the bookshelves [albums of comics] . . . settling in even he is only able to cover his head over hindquarters . . . —responding to the feline need for seeking refuge or hiding in order to sleep or observe.

Dewey becomes enamored with one such specific area of the library where, at the brightest hour of the day (late morning), he spawls and leaps up onto a gigantic surface (a large desk), gathering himself before pawing and battling it out with various items, which he sends flying into the air (plastic protective sleeves for covering the books). After some time has passed (1989; one year) he discovers another means (a ladder) for climbing to a higher point in the library (all the way to the top of the bookshelves). Here. Dewey can reign down over the area and observe, resting there for a while during the brightest hours of the day, or setting off and leaping onto ever higher items (such as a row of lights hanging overhead) so as to observe without being seen. If he eventually sees a portion of his various asylums vanish (*due to the initial technology update library in 1994, six years after his arrival*) he quickly adopts new objects to lay on for his daily naps. These new objects don't necessarily protect him but nevertheless keep him warm (*computers, which would render card catalogues obsolete*), similar to other objects (such as the photocopiers) he would spread out across due to their emanating warm air. No doubt, he made use of his territory in a completely different fashion when left alone at night or on days left in the silence and brilliance of the morning light (*on Sundays*). For it's on Sundays that Dewey learned how to drink water from the toilets and subsequently will eventually refuse to drink from the bowls offered by the library workers.[4]

Nevertheless, the library doesn't exactly function like an ecological nest since he is fed and doesn't depend on the library for survival. Rather, the library becomes something of an inviting space for him to adapt to his personality and felinity, which he transforms and reconfigures into his territory, henceforth bearing his mark: *the personnel and the public no longer see the space as a library and no longer evolve in the same way, from then on paying more attention to other aspects of the space that had previously been ignored*. And all of this interaction slowly gives shape to him: Dewey is no longer the same once he has mastered his new environment, which takes about a year or so, and he suddenly begins to display himself differently after conquering these various summits.[5]

We also see that the later transformation of his territory (the library's renovation in the summer of 1989) disturbs him to the point that he flees. *According to Vicki Myron*, during this period, (a three week stay at her home during the library's renovation) Dewey would be able to stare out the window. To see and feel the birds along with the wind. But this wasn't Dewey's first vacation. More than likely, upon his return, he notices another sort of luminosity (*the walls are now painted and are no longer the dull grey of concrete; the floor is now blue and no longer orange; there is new colored furniture*), another space (*tables, chairs, and bookshelves that have been rearranged*), other scents (*a new carpet and books*), "his entire universe had changed." Having become "jumpy," he constantly lounges near the library entrance, which he used to ignore, attempting to leave, no longer heeding to the calls by the staff and eventually escaping (after about eight days), perhaps in search of his lost old territory. He rediscovers an odor, a form, a familiar sound (*one of the library workers will discover him three days later underneath a car*), and there he lies, "hunched against a wheel of a car," huddled and covered in motor oil, with a torn ear and a scratch on his nose, trembling, no doubt frightened by the constant rumblings of car engines and the anger of his peers. He lets himself be picked up and immediately purrs and cuddles against the human he knows the best (Vicki). He accepts being wet (another bath in the sink) in order to remove the stench and motor oil, and henceforth no longer tries to escape even when strange winds and odors envelop him from the outside signaling to him that the library entrance is open. And conversely seeks out a hiding spot in particular if he hears the rumbling of a passing truck. "Dewey was completely done with the outdoors."[6]

And yet, Dewey doesn't exactly become a member of the territorial culture of inside cats. He needs humans and his sociability determines his demeanor. Dewey demonstrates this immediately upon his arrival, and the neutering he undergoes not too long after cannot explain everything in regard to his disposition even if it perhaps reinforces it. As such, he constructs his territory through a series of encounters. Each morning, when the light begins to shine through the front entrance of the library, having probably learned to recognize the time of day through the strength of scent, the intensity of the luminosity and his internal biological clock, Dewey waits patiently to welcome, with a careful observation, every regular human he knows (mostly the library staff). "At two minutes to nine, Dewey would drop everything and race for the front door." Waiting for the intermittent arrival of regular patrons. The rest of the time, he would run back to the front when regular groups of younger patrons arrived (a class of students or those with disabilities) or groups of older patrons (participants in book clubs). With the former, he wEAvEs his WAy, ruBbing AGaInst ThEm, clIMBInG on One OF them, never the same one. With the latter, he lEaPS UP onto the specific height in question (the table), MAKEs HIS WAY AROUND those PRESENT, QUIVER-SNIFFing THEIR hand or head, thus bringing humans into contact with the most pertinent zones for cats, detecting and recognizing odors, perhaps even pheromones, and forms, chosen through SITTING on their laps. As for the other older adults, Dewey will come running upon the arrival of any unknown humans, whose numbers will continue to rise (in parallel with his increasing renown and popularity throughout the rest of the United States, with some people traveling several hours in order to specially pay him a visit), which he will then see surround him, staring at him, petting him, and then leaving again—all of whom he will willingly spend time with (and also time for a photograph)—or lets himself be picked up, *mostly by children*, not squirming one bit even if he finds himself turned upside down, with his head below and his tail up in the air.

He often strives for attention. He hops up into the laps of part-time workers, continuing to beg for their attention if he is placed back down on the ground or moves on to the next worker if no interest is shown. He will observe everything from the height of a lap or rolls himself up into a ball and naps, thus sleeping throughout the day in brief parcels of time, slumbering in what is certainly a very light sleep. He more than likely is able to regain a certain mild warmth after the cold nights spent with the heat turned off. But he also does the same thing when it's hot outside and so his actions instead reveal something of a quest for another objective: a reassuring and soothing contact. And it's no doubt for the very same reason that he seeks out nearby objects (papers and binders) that also bear the workers' scent. As for the other full-time staff (besides Vicki), he makes his way from one to the other choosing them in accordance with their emotional state of being, something cats can do once they've come to familiarize themselves well with their humans and decipher them, rubbing against their legs in order to exchange scents, revivifying the common space, expressing their pleasure, and professing their friendly intentions. He lEAPs up to their same level of height (onto their desks), tries again if he doesn't quite make it up on the first jump, settling down on nearby objects (books, papers, and files) TUrNs his back to the human, thereby transforming a bodily technique into a ritualized behavior that is slowly but surely perfected, more than likely in order to elicit interactions.[7] Equipped with a specific cognitive and sensorial framework, with a psychological singularity, and with an individual agenda, Dewey co-constructs his ways of being with this particular environment, fashioning behaviors just as particular.

While he sometimes receives rejections, most often Dewey is welcomed with caresses, curling up in a ball or turning his back or head against the human, which he prefers the most as is common with many cats since they also do this with each other, whereas he shows his belly if he is repulsed. And this is how he makes himself understood by the humans who are closest to him, *those who spend the most time in his presence learning to identify his preferences as is the case now with a number of cat owners today . . . in the West*. Dewey more than likely solicits petting in order to exchange scents, perhaps pheromones, thus reinforcing his inclusion into the community in which he must think of himself as a member, reassuring himself of his ties and feeling a certain pleasure. He receives these tactile messages from the humans due to the changing conception and relations with felines that took place over several generations in the West. In particular the changing relations with common cats such as Dewey, to such an extent that these modalities of interaction, with a cat seeking out human caresses and the humans welcoming them, are now included in a number of feline education manuals. And Dewey has an advantage since he has both the possibility and the penchant for moving from one human to another so as to erase any refusals or lassitude, thus constantly affirming and reenforcing the social connection and subsequently prolonging his well-being.

He also solicits and obtains game play but only reserves these activities for the permanent staff, those he knows best: climbing up to their level (the desk), casting objects with a swipe of his paws. He seeks out easily rolled items (pens and pencils) that he then catches and sets in motion again. This allows Dewey to reinvest his aptitude for hunting (which he will never use) toward nonexistent prey. His humans will nevertheless still look on. Sometimes he will simply push the items with his head and watch them fall off the desk to the floor. He feels comfortable in perceiving the cries from humans (laughter), and other movements, vibrations, and scents that express a certain pleasure on their part, which Dewey more than likely deciphers, since we know that cats have the ability to acquire the faculty of reading human emotion—a skill that is not innate but constructed. He frequently receives prolonged forms of reciprocal satisfaction in the form of a ball (of crumpled paper) tossed from a distance, for which he casts out his paws, stirring and tossing them in order to claw the object, quiver-sniff it, abandon it, and then begin again as soon as he sees it tossed again.[8]

> You must be saying to yourself that you have partaken in these very same experiences with your cat! Certainly that is true, since Dewey and his humans embody current exchanges that are more frequent than in previous eras. They rest on the acquisition of a reciprocal knowledge: the adoption of the other's viewpoint and a reinforcement of interspecies connections, of a conviviality, and a conjugated well-being. Neurobiologists and ethologists have proven that both species share a reciprocal desire to engage, founded upon homologous brain functions and structures for social behavior and the resulting emotions. They both share in the same adaptive individual decision-making process and the mesolimbic reward pathway. The scientists could have just as well referred to "social physiology" and the "social brain." This system allows for an empathic comprehension of other individuals of the same species but also of other species equipped with similar mechanisms of behavioral expression. Following along through these pages, you have perhaps come to realize that such a biological con-

> dition is obviously necessary but not sufficient. Mutual will also depends on historical, environmental, social, cultural, and psychological factors on the part of both cats and humans. And this is also a construction. It is not independent of the biological, which it needs in order to be expressed and concretely embodied. And this mutual will demonstrates, here as much as elsewhere, a dynamic entanglement, going from the biological to the cultural.[9]

Dewey mobilizes and reroutes his feline passion for rapid movements, that is, signs of potential prey, which also lead him to take note of other buzzing and whirring sounds (typewriters), luminous flashes (the photocopier), and colored vibrations (film projections). "Then he would jump on a table, curl his legs under his body, and watch the screen intently" alongside *the children*. And Dewey goes to great lengths to retain this sense of cooperation. For example, through selecting specific body signals and adapting them—remaining at a specific desired spot, meowing while staring at the human associated with it—he is able to get the person to *regularly make flowing water appear* (from the sink), which he doesn't drink but contemplates falling and disappearing, "bouncing off the drain plug," sometimes "taking a quick slap at it with his paw." As far as the employees are concerned, they receive pleasure through watching him frolic about and have fun with him. They bring him things to bat around like a red ball of yarn dangled high above his head which he then tries to knock down, or various other toys initially proposed by consumer companies and eventually requested and adopted by cat owners for everyday use, items that would become normalized during the second half of the twentieth century: a ball he runs after; a stuffed mouse which he casts out, then catches, and tosses out again.

Thanks to such a complicity, Dewey experiences nothing less than vegetal euphoria! He immediately notices whenever one of the library employees arrives with catnip, who rejoice in inciting what becomes referred to as the "Dewey Mambo":— . . . QUIvEr-SNiFf . . . scratch teethes chEWs . . . CoNVulSEs COnvULSEs . . . RoLLs around on the ground rOLLs . . . *twisting* and CRAWLING mOVinG his PaWs . . . stopping . . . EXHAuStEd . . . — like other cats who are reactive to these herbs due to a genetic allele making them sensitive to the molecules emitted. He experiences the same sensation each time during the dark and cold period with the appearance of a great odiferous item (an artificial Christmas Tree). Every time he does the same thing:—QuIVEr-SniFf EaR-tUrNs pupillAtes . . . surPriSED ShAkEn with Emotion . . . tossing out, moving and planting his paws . . . HEaD EaRS WHIskErs PoinTeD TAiL . . . QUIvEr-SniFFs for quite a while above his head (boxes and branches) . . . CLaWing and ChEWing any fallen strands . . . FrOLIcking FRoLicking around . . . plopping down in front of it, remaining there quite a while . . . QUiVeR-SnIfFing QUiVeR-SnIfFing . . . *giving the impression that he as well is enjoying the holiday festivities*. His passion for these specific scents leads him to quickly recognize them, begging with insistence, taking advantage of any absence or inattention to return to the items and revealing through a refined observation a good overall awareness and sense of humans, getting caught up in the ruses of tiny soft objects with strong odors (rubber bands), which he chews on while bounding about, sometimes even swallowing them. Their rarefication, even their disappearance (*in order to prevent him from swallowing them*), leads to nighttime excursions:—pushing his paws from one site to another; attentively quiver-sniffing; locating areas traces of lingering perfume; jumping and climbing up onto higher surfaces (the desks); clawing and chewing and chiseling on the tense smooth layers attached to objects (cards and documents); introducing and weaving his paws penetrating into any spot with lingering perfume enveloped in pleasant odors (drawers and closets) quiver-sniffed through the openings or located by watching humans . . . — Together, they all give themselves over to play, understood by each party, which he will abandon for a while as he grows older, and which he will return to enjoying five years later.

We should note that Dewey shows himself to be much more interactive than many other library cats. He embodies a phenomenon highlighted by ethologists and veterinarians starting in the 1950s first for dogs and then for cats: the increasing persistence of so-called childish behaviors, notably that of the play. Based on the fact that these animals are now no longer beholden to official work and are better cared for and find their company solicited in a new manner: there is a much more intense, ludic, emotional, and interactive engagement with adults or children who are all the more drawn to such behaviors. And once again, the term "childish" should be carefully reflected upon. It was evoked for a situation judged

> to be normal or natural, but which in reality is purely historical: cats and dogs prior to the 1950s only played as puppies and kittens, and most often among themselves, rarely with humans. Rather than viewing the behavioral transformation as the persistence of childhood, suggesting some kind of shift in the entire species, it would perhaps be better to think in terms of a profound change in specific individual canines or felines. Following the same paths as Western humans themselves, who begin to work less, also having attained better living conditions, becoming themselves more and more interactive among themselves, becoming players, athletes, and thus more interactive with their own children . . . and also . . . with their pets.

Dewey also co-constructs other original interactions that are fairly new within the history of cats. Having noticed that the book cart moves, he becomes accustomed to hopping up onto the *cart waiting for when the cart is full of books until one of his attentive humans* notices his desire to move, and readily pushes him on it. "Seated on the cart, he would look over the rows of books as they passed by and when something pleasing caught his eye, he would indicate to Joy that he wanted to get down, as if he were seated in a shopping cart for cats." With another attentive female employee (women made up most of the workforce, also showing themselves to be more attentive and interactive), he connives his way into being allowed to sit on her left shoulder, "never the right," and to accompany her, both of them creating what will be referred to and adopted by many of the staff as the "Dewey Carry."[10]

Nevertheless, he still consistently maintains the most singular and strongest relationship with *Vicki*. More than likely, he retains a memory of his initial experience, which he will relive again not too long afterward during another moment of fatigue and stress (his neutering at the vet clinic), *when Vicki reappears and picks him up*. Reinforced by this new state, he is still affectionate: voluntarily sitting on her in order to mix their scents together, reaffirming their attachment, expressing his well-being by purring; accepting poses that are hardly feline and much more human-like, such as being cradled against her chest, his head on her shoulder, against her neck, more than likely because he appreciates *the vibrations, odors, and perhaps even the pheromones of human pleasure*. And through such contact, he will transmit his own gestures and postures, signs of a searching and accepting that will be well received by his human.

To the extent that Vicki was available—not so much that she was isolated but when she wasn't busy taking care of her daughter, or helping other friends, relatives, or readers, and she was so inclined—she is attentive to Dewey who she deciphers rather well; she appreciates his ongoing quests, to which she willingly responds, and their reciprocal "empathy," whereby each of them guesses correctly the expectations of the other thanks to nonverbal communication, comprised of signs, that is, postures, gestures, vibrations, and emanations. She takes to "loving" this cat, experiencing a love for this cat who "treated people so well," "intelligent," "playful," and therefore responsive, and she integrates this affection within a framework of a child-to-his-mother, considering Dewey as her "baby," exclaiming to him: "I'm your mama, aren't I?," obviously in a metaphorical manner, as far as her understanding of taking on a responsibility, sympathy, and an engaged attachment.[11]

> Vicki embodies a new generation of cat owners: those desiring a considerate companion, much different than those owners of the nineteenth and twentieth century, who merely appreciated a loose, distant, and independent company with their cats. And Dewey represents a new type of cat that will become more and more sought after since the last quarter of the twentieth century. The criteria of friendliness, sociability, reactivity, intelligence, and a playful spirit all become important when choosing a cat among the breeders, those who sell cats, or those found in shelters, the latter movement having first begun in the Anglo-Saxon countries before spreading throughout the rest of Western Europe. To such an extent that those working in shelters place children's toys into the cages where the felines are kept so as to promote a taste for play and sociability in order to increase their chances of adoption.
>
> Parallel to this, recent documents and studies have demonstrated a growing desire for owners to understand their cats. What was once a desire reserved for a small few has now become an obligation. Contemporary cat owners seek to be attentive, sweet, affectionate, and prepared to adapt themselves to their cats. They speak with their cats and therefore experience more interactions, even soliciting them, most notably in the interaction of play, which was also once a marginal activity. However, today, playing with one's cat is considered of vital importance and even suggested as necessary in the very education of the feline in various books on rais-

> ing a cat. Such educational manuals are often inspired by or heralding from Anglo-Saxon countries, who also privilege a more comprehensive approach. All of them following a contemporary model initially created for children but later adapted to cater to dogs.
> This type of cat-child is not a mere case of a projection of human desire, nor is it gratuitous. It constitutes at once the engine of a dynamics and its product, since it has recently been shown that such a dynamic gives way to more attentive humans and more reactive cats, through mutual collisions, that is, that such a reality influences both cats and humans. And far from being a universal situation, this is still a very recent phenomena, but it perhaps can serve as an important indication for orienting the future.[12]

Dewey reinforces these kinds of connections by establishing them with Vicki and then repeating encounters and gestures that are quickly codified, functioning as rites. Early on (*during the first week upon his arrival in the library*), he notes she is always the first to arrive in conjunction with the return of the bright luminosity of the morning. Dewey waits for Vicki at the library entrance and then runs to his food bowl to demonstrate his hunger, and then follows her throughout the rest of the territory, listening to her *call for him*. This exchange evolves (1997) when Vicki begins to carry him to his litter box, suspecting an issue of constant constipation. And soon after, he will no longer go to the bathroom in the litter box without first snuggling next to Vicki, jumping up into her lap, asking to be cuddled. Later on, after a long period of Vicki's absence (due to breast cancer, 1998–1999), Dewey suddenly begins to "wave at her" when Vicki would arrive into the parking lot, scratching his left paw on the front door, continuing to do so as she crossed the street and entered the library. And Dewey will continue to repeat this daily gesture until his death, which Vicki interprets as his way of welcoming her. However, these two readings, one egocentric and its opposite, don't so much cancel each other out but align themselves.

As with the various games:—scurrying off to fetch *tossed* balls; or, right before her disappearance [each day in the late afternoon] and after imposing his presence and his expectations by sitting in front of *her*, on her object [the keyboard], running with her, separated by a row of shelves [*running from one row of books to another*], scurrying, moving, stopping together and heading off again in the opposite direction. Or when, after several sounds made by Vicki (*always the same*), having been spotted by her, he darts off to play hide-and-seek between the books. Waiting for *her to approach* before darting off to hide again . . . — displaying here again a very subtle knowledge and recognition of *the other*, which allows him to anticipate, respond, and react, without which none of his actions would be possible. Vicki is well aware of this reciprocal comprehension. She judges such comprehension to be stronger on the part of Dewey, who she claims, "always *seemed to know what I wanted*," more than likely because he always read her on several levels by way of supplementary signs (sounds, sighs, gestures, and odors . . .). It felt good for Dewey to obey, even if he would flee *whenever she uttered the word "brush" or "bath"*![13]

Moreover, Dewey only ever leaves the library *with Vicki* and never in a cage. She wants to avoid reminding Dewey of his traumatic time spent stuck in the library drop box. And every time he leaves with her, he is wrapped in a warm cloth (a green towel), whose odor, texture, and color alert him to his immediate departure. Such excursions are frequent enough that he is able to imprint a sense of them (an orientation, the distance, landmarks) throughout a long jolting and noisy moment to which he reacts accordingly: crouching on the floor in the backseat of the car behind his human as long as the directional path appears to be headed toward a dreaded location (the veterinarian clinic); hOpPing up onto higher surfaces to look outside when he finally realizes he's headed toward *Vicki's house, during prolonged closures of the library over the holidays*.

He jumps as soon as he detects something novel (an open door), GaL-Loping toward this site, gathering information through his olfactory sense, regaining possession of this more reduced territory (one thousand square feet), one less diversified (with just three rooms and two humans). He quiver-sniffs in every direction upon his arrival, including throughout the first couple of nights, climbing and hurtling himself up and down inclines [stairs] that take him to a cold and dark space (the basement), an area he has never experienced before at the library. Then, he hops up onto a favorite elevated surface level (for example, the back of the sofa), carefully examining the outside where he is not allowed to go. *Vicki, being fearful of Dewey running off*, adopts an attitude of Dewey as a house cat, above all when he is left alone (during grocery shopping, holiday parties, etc.) Otherwise, Dewey always places himself next to his human (on the sofa), including at night when he hops (up onto the bed) and cuddles up close, rolling himself into a ball, another difference from his normal daily routine whereby he can lounge for long durations, mixing scents with Vicki, further cementing their connection.

He becomes even more attached to the other human in the house (Vicki's teenage daughter, Jodi): waiting for her to come home from school, watching her, running to her side at the first sight of her (at the door) or as soon as he hears her at a distance or glimpses her, following her everywhere "like a dog" (even into the bathroom), and if he's not let in, he waits on the other side of the barrier between them (the other side of the bathroom door), meowing to get her to return, climbing up onto her lap as soon as the occasion arises (when she sits down), rubbing his head against her and "purr[ring], purr[ing], purr[ing]," jumping up beside her at night (in the bed) attaching to her like "velcro." He continues to sleep in this same space full of her scent even after she has left (returning to high school), jumping up and down around Jodi upon her return, sticking to her for contact, and later (2005) he will eventually transfer this attachment to several new humans with a similar scent (Jodi's twin daughters), not following them everywhere (due to his old age) but allowing himself to be petted (on the back and head), thus adapting his world to the changing environment and constantly (re)constructing it.

Nevertheless, his attachment is not exclusive to one person. When he is in his own territory at the library, he can stay for a while without seeking out his human, brushing against others, which he also does during another long period of time (*when Vicki has cancer*). Experiencing a number of daily contacts with staff and patrons as well as partaking in numerous activities, Dewey doesn't appear to live the common life of an indoor cat: he doesn't often sit and simply stare out the window, he doesn't become obese or aggressive, or contract any form of adaptative illness besides bouts of constipation, which lead him to often change the types of cat food he eats and be carried to his litter box.[14]

Conversely, from the very first days of his adoption, Dewey increasingly influences humans in concert with his ever-increasing reputation. Vicki notices that the adults solicited by Dewey tend to return more often to the library, staying for longer periods of time, engaging more with the library staff in order to talk about the cat. His attraction has an even stronger influence on the children, who beg for him to come to them, seeking him out as soon as they step foot into the library, dedicating their time there to Dewey. Kindergartners get excited upon seeing him but calm down so they can pet him. Young disabled children derive a certain relief from him and are fascinated by his presence. Dewey exerts an overall calm and emotional support, which has recently been understood in the West and other Westernized countries as the therapeutic capacity contained within animals such as cats, dogs, and horses . . . , far too often forgetting that what they are witnessing is a historical situation. If Westerners no longer faint when coming into contact with a cat but actually receive a positive increase in their mood, it's because they have changed . . . just like the cats! As such, on the arrival of his first estimated birthday, cards, drawings, and well-wishes abound for Dewey. And when the day finally arrives, the patrons along with excited children and their parents pour into the library to watch Dewey weave around them and taste his special cake made in the shape of . . . a mouse, even though he has hardly ever encountered one! And the following first Christmas spent at the library, Dewey receives an assortment of balls, treats, and plastic mice.[15]

> And this situation gives us an idea of why this cat collective is so interesting. It reveals a certain state of Western culture between the twentieth and twenty-first centuries, prepared to accept, integrate, and collectively live with the status of the cat as friend, which was not the case during the time of Marie Dormoy for instance, but which will eventually become normalized, as was similar with dogs between the nineteenth and twentieth centuries. In parallel to this shift in the human status toward cats, Dewey also displays a new status of the cat as child, that is, the cat as a member of the family, following a similar shift for dogs in the second half of the twentieth century.
>
> In the end, Dewey reveals one of the modalities of the social diffusion of such new types of relations of the "cat as child" through the various media outlets that immediately gravitate toward him as soon as he is adopted by the library. First, it's the regional press that facilitates this status and then amplifies its acceptance even though a large portion of the local population still only views cats through the model of a farm cat. Then, the national press, along with radio and television stations, latch on to Dewey's story, transforming him into a celebrity and subsequently drawing visitors from the four corners of the nation to Spencer, Iowa, thereby diffusing a certain type of cat and model of relations. This follows precisely the pattern for dogs at the end of the nineteenth century through the first half of the twentieth, from Greyfriars Bobby in England to Baushan in Germany, and Rintintin in the United States. To this list of famous family dogs we should also add several fictional dogs invented in the twentieth century through literature, cinema, cartoons, and comic strips. All of which played an essential role in establishing the status of the dog as friend and later on, the status of dog as child. Fictional dogs such as Lassie, Milou, Dogmatix,

> Bill, now all of them in competition with such fictional cats who are just as close to humans, from Garfield to the star of the more recent twenty-first-century French film, *The Rabbi's Cat*. In the end, as we moved into the twenty-first century, this type of cat similar to Dewey—affectionate, sociable, and interactive—has now largely been diffused thanks to the internet by way of short, shared video clips posted by the cats' owners.[16]

As he grows older (starting around the year 2000), Dewey slowly begins to change his habits:—he no longer climbs up to the highest peaks [above the bookshelves and to the dangling ceiling lights] . . . tending to reduce the amount of pressure he puts on his painful paws . . . meows asking for others to carry him and set him down on the book cart . . . is now "cautious around toddlers," remaining more still, spread out like a Sphinx in order to perceive, sniff, and examine his surroundings, . . . still privileging the library entrance in order to welcome the library patrons, perched from on high [atop a desk] sheltered from some of their friendly gestures that would now potentially harm the "old grandpa Dew" . . . protected by one of the employees of the library who works at the welcome desk and who makes the suggestion that it is time to take care of the old "grandpa." Then, he becomes deaf, struggles to clean himself, limps, suffers from arthritis, and begins to have even more trouble eating his food. He forces the library staff to adapt to his changing condition, to give him more attention and time along with more money in veterinarian costs and food, which he happily accepts to the extent that his dependence on humans has by now become normal and this cat friend or child is now inscribed within this same community.

But Dewey never displays any signs of aggression, even when he begins to express (at the age of eighteen, beginning in 2006) signs of suffering: unable to control when he urinates, meowing and limping because he has developed a malignant tumor in his stomach. Dewey's suffering and diagnosis lead Vicki to make the decision of having him euthanized. He is cremated and his ashes are buried on library property. His death is announced by the national media, which unleashes a wave of sympathy; the townspeople of Spencer organize a wake at the library. Vicki suffers "a horrible shock" and tries to overcome her mourning by writing a book about Dewey.

Here again, Dewey is a good depiction for other pet cats of our contemporary era. He embodies not the first appearance (which happened much earlier) but the social normalization of the old cat, which began during the second half of the twentieth century, most notably thanks to an abundant source of food and ongoing improvements for the healthcare of cats, to such a degree that there is a sharp increase in veterinarians specializing in pet cats, following a similar increase for dogs, who seek to provide preventative and regenerative care so as to prolong the relationship. For these cats who enjoy a new status and new ways of being are now precious to their humans, who want to and now are able to spend more money on them as a result of the increase in wealth in the West. Whereas in the past, cat owners would quickly get rid of their cats or dogs who suddenly began to become useless to them—through drowning, hanging, or clubbing them. These much older contemporary cats now experience similar troubles of aging like their human counterparts: from arthritis to obesity, from incontinence to anorexia, from hearing loss to senility, from passivity to aggression.[17]

Dewey also embodies the increase in lifespan of cats, following in the footsteps of both humans and dogs. A longevity that first is established in more wealthy families as in the case of Bibiche and Miton but will become widespread in the second half of the twentieth century. Meanwhile stray cats and farm cats will still continue to experience a rather short lifespan, on average, of about three to five years. Dewey eventually dies of cancer, one of the most common ways for both cats and dogs to now die, not to mention for humans, since they all share in the same environment and certain illnesses also are spread in a parallel way (a decline in rickets and viral or parasitic infections, and an in increase cardiovascular and cancerous pathologies).[18]

The funeral arrangements for Dewey are also symbolic of another growing trend. The burial of pets on the property owners' land was already established by the aristocracy between the fifteenth and eighteenth century, adopted thereafter by the bourgeoisie, as was the case for Moumoutte and Miton, and then popularized throughout the twentieth century, with burial in the middle-class gardens slowly replacing the old way of simply tossing their carcasses into ditches, holes, rivers, or streets. Since the end of the twentieth century, the cremation of pets has continued to increase, now becoming almost the only way in which the bodies of pet cats and dogs are disposed of, following the same tendency in urban areas for humans beginning in the early 1980s. And the mourning of humans for cats became commonplace during the last third of the twentieth century, following a similar development for dogs. What were once marginal emotional states expressed by a small minority such as Marie Dormoy and

> Athénaïs Michelet have now become widespread and normalized, or at the least, have now become more discussed, the mourning phase becoming more and more intense it would seem, the very proclamation reinforcing the profound feelings and vice versa.[19]
> This model of the cat-child, of the cat as a family member, seems to have emerged and is still the most represented in Anglo-Saxon countries, but it is also beginning to take place more often in Western Europe. For the moment, such a shift in cat representation still seems to be less prevalent in other Western European countries but it's difficult to estimate the exact numbers. Readers' letters, podcasts, and other short videos sent to television shows or posted online within the increasing speed of immediate circulation continuously show how this new relation to cats is evolving. We can also get an idea of the increase in such a relationship between the human and the cat-child by way of the continued success of translations of Anglo-Saxon biographies that promoted and spread this model. As such, countries such as Spain have quickly seen an increase in the number of foreign bestsellers in translation at the beginning of the twenty-first century about cats, promoting the notion of the cat-child, which had hitherto been nonexistent. Such firsthand accounts are still rare in Europe while others are fairly significant, such as that written by Maria Rosenkränzer regarding her cat Lilly—a tabby cat in the suburbs of Frankfurt from 1997–1998.

Lilly is between the age of five and eight months when she shows up to rub *against the legs of Maria's daughter along a forest path*, no doubt lost or abandoned and begging to be adopted. She immediately shows herself to be sociable, affectionate, and interactive. She quickly warms to the apartment, constantly leaping up on their laps, playing with Maria and her children, playing hide-and-seek, or joining them in the kitchen. These humans willingly and cheerfully accept her but are also surprised, having never experienced such a conviviality with either of their two other cats at the same age. The rupture of such human behavior is also occasioned by Lilly's manners with the other cats, who moreover rarely mixes it up with her two other older peers (both sixteen years old) and who are above all independent and bearers of another culture. Lilly sleeps in the beds of her humans, participates in dinners around the table, is part of the Christmas holiday and receives a gift, tolerates being dressed up for play, and even enjoys being in the shower. She is also taught to not bother the birds or the guinea pig in their respective cages. And even on the rare occasion she is let outside, she has been taught not to hunt. But she eventually escapes from the yard and gets hit by a car, *dying in Maria's arms, as she whispers in her ear. Although it was a very brief moment (just nine months) the experience made such an impact on her family that Maria decides to write about Lilly in order to evoke this "person."*[20]

The diffusion of anthropozized feline cultures is the result of two processes. The first is that of humans, which was extended all the way to welcoming stray cats. Such was the case with Dewey and certainly was the case with the cat Bob who was adopted in London in 2007 by James Bowen, a nonconformist who became an acrobat and eventually a street busker, and who writes about the cat in a bestseller published in 2012:[21] the book sold over a million copies in Great Britain and was eventually translated into thirty languages, in addition to being made into a major motion picture in 2016.

This street cat, struggling and starving, while already being a mouse-hunter and scavenger of the trash cans, quickly shows himself to be sensitive to *James's attention, who had been around cats since his early childhood, speaking to Bob, petting him, feeding him, and naming him.* He lets himself be drawn into James's apartment at night, accepts food and care, develops interactions, and above all patiently waits by the building during the day *when James is busking in the streets.* The cat begins, however, to display a significant change in behavior once he begins to accompany James around, merging the capacity of a street cat to wander about with the attachment of a cat-child, moving from a territorial culture to an anthropozized one. Bob's demeanor becomes all the more interesting and sensitive once he adapts to wearing a leash and then a harness, *James fearing he might lose him as Bob begins to become a public attraction through which he makes money.* Bob even accepts wearing a winter sweater, then forgoes getting into trash cans after falling ill and begins to adopt human manners while in the apartment, even learning how to make use of the human bathroom.

The human dynamic of pressure also has an effect on groups of stray cats, in particular those living in an animal shelter. Established in various large cities in the United States and Great Britain, as well as France, in the late nineteenth century to welcome cats who had been chased off or other inside cats who had been abandoned, these shelters began to multiply in particular after the social success of the idea of the pet cat, which subsequently led to even more abandoned cats.[22] Such shelters have even given rise to a specific ethology and veterinarians specialized in shelter animals beginning in the early 2000s, with multiple studies on stress in shelter animals, their management, and possible substitutes to euthanasia, which is often conducted on a large number of animals. Such human intervention has sometimes even spread to ways of being, and the old form of nonintervention has now been replaced by a desire for order. Let's take two extreme examples from the twentieth century with contrasting intentions, the first being chosen due to its reliable account on the one hand, the other thanks to a large study, which is rare. While neither of them herald from the same country of origin, these studies demonstrate nevertheless that the dynamics at work are analogous for both territories.

The first example is provided by Paul Léautaud, who welcomes both cats and dogs into his abode beginning in 1912 when he moved to Fontenay-aux-Roses, where he will transform his land into a shelter, housing anywhere between twenty and forty cats during World War II. Guided by the notion of cats being naturally free and independent, Léautaud only provides them with shelter and food, never intervening into feline ways of being, allowing various sociabilities to establish themselves, at the level of the group, couple, of the adopted adult, and kitten, cat and dog, with varying empathies and discords along with solitary and sociable personalities. He also lets life happen: from reproduction to deadly contagious illnesses, including cats being killed by dogs. And given that he lets the cats mix with him as they wish, some of them settle inside his house, others run off for long periods, and some simply hover around the margins.[23] But do not be fooled. Contrary to what Léautaud thinks, none of these ways of being are truly natural, but rather, they are adapted according to the environment he provided.

In complete contrast to this, the recent animal shelter built in a house in a neighborhood in Albany, New York, magnificently studied by two socio-zoologists in the early 1990s,[24] is managed according to other rules by other animal protectors nourished by a different representation of cats, other animals, and life in general.

Furthermore, the individuals running the shelter are also against the common notion of solitary, aggressive cats, or allowing them to fall into living in some kind of hierarchy. And in that light, feline competition and violence are prohibited in favor of a friendly, social egalitarian organization of a culture of tolerance and sharing, and there is a huge array of concrete arrangements so as to maintain such an order (at the level of cages, rewards, punishments, etc.). And since the cats are never euthanized and are allowed the time to adapt, they eventually accept and respond to the structure, creating a social cohesion around some of them (certain individuals

whose task it is to socialize new arrivals into the shelter or those who are clearly viewed more as friends), living within a "cooperative culture" with adjusted personalities, even if there is a clear separation between the newcomers who are still disturbed and adapting and the older, adapted ones, between the cats who were originally vagabonds and are quick to adhere to the new social organization of the cats and the former cats who were expelled from the living rooms of the bourgeois and who feel closer to the humans than to the other felines and who are therefore more stubborn. And here again, one mustn't be misled: the cats are free to express their social preference because the humans in charge of this shelter hold fast to Épinal's image of cats, but this autonomy takes place within a very anthroposized framework and with a very suggestive orientation.
And yet, such a dynamic pressure by humans doesn't prevent a second impetus for the diffusion of anthroposized feline cultures by the very cats themselves, who solicit the humans' attention, taking up initiatives for such engagement, as we saw with Dewey, Lilly, or Bob. These cats beg more for human interaction than their predecessors or even contemporary cats, at times even behaving like dogs, a tendency we see even more accentuated with . . .

CHAPTER 8

Catdogs

. . . who embody the most recent level of anthroposization and, for the moment, the most advanced. As a result of the drop-off in the rates of dog owners in Western countries (the South Pacific, North America, and Western Europe) since the end of the twentieth century and the strong increase in cat ownership, a significant percentage of cat owners adopt cats for different reasons than did owners between the 1960s–1990s. No longer is it a question of countering show dog owners, rather, their choice is a decision to replace dogs, who they deem to be cumbersome, with a more practical choice of a cat while simultaneously placing on the feline's shoulders their desires for proximity and companionship, the types and modes of relations that were previously applied to the dog. These humans prefer and give precedence to more interactive felines, those who like to remain in close proximity and who are more connected to them, and who are increasingly becoming more and more like pet dogs. Such felines are part of the process of becoming what I refer to as "catdogs." The best account we have so far of this contemporary trend is about . . .

Jonah: The Anxious One

(Melbourne, 2008–2012)

. . . evoked by the New Zealander Hellen Brown, born in 1954, who later remarried but retained the last name of the father of her children. She is a member of an urban upper-middle class family, works as an independent journalist, and is the author of over a dozen books on various subjects. She enjoyed an enormous success with the publication of *Cleo* (2009, two million copies sold of the English language edition, and translated into seventeen languages), a book that tells the story of a cat who heals the wounds of a family after the accidental death of their young son, Sam.[1] It is a book she writes following the death of Cleo, in Melbourne during the very same time she also had the cat Jonah. Swept up in the accolades for her first cat book, and even though he is still a very young cat, she ends up writing a second book—*After Cleo Came Jonah*—in 2012.[2] Moreover, at the time I write this book, Jonah is still alive.

While simply focusing on the first years of Jonah's life, the work allows us to construct a portrait of a catdog, marked by separation anxiety with his owners, which is a rare occurrence in felines but which some cats have begun to experience following in the footsteps of dogs, who were initially considered pet companions, then friends, and finally as children. Helen Brown's account is all the more interesting given that the rise of the catdog is a phenomenon that arose out of the Anglo-Saxon countries of New Zealand and Australia, due to these countries' particular environment, in which people sought to move cats inside and eradicate those that remained outdoors.

In reality, the two books are less about the cats than the recounting of a number of personal and familial difficulties (the mourning for her son, a divorce, breast cancer, the Buddhist inclinations of a daughter . . .). However, the cats stand out as a psychological counterbalance helping to provide comfort and assuring a larger family cohesion similar to what the bourgeoisie and its literature expected from dogs during the nineteenth and twentieth centuries. And the cats are well portrayed and described, in particular Jonah, who is talked about in a more transparent manner than Cleo. We also get a good picture of the psychological and environmental evolution of the family as they go from accepting one kind of cat to embracing another, accepting him and encouraging him.

We must, however, be prudent in deciphering the literary performativity and staging at work. For example, we see the desire to connect human stories with feline ones emphasizing certain traits: claiming Cleo is more attached to Helen's family than is really the case (a cat who would sleep each night on their roof) in order to help them through the mourning process.

Or claiming that Jonah is more independent than he really is (when he appears to be very attached to the family) so as to create a parallel between him and one of Helen's daughters. We must also pay careful attention to certain accidental shifts or even conflations of one cat with the other, more than likely due to the fact that the book on Cleo was edited at the same time as the book about Jonah and that both books retain the very same narrative structure and composition style. We see a certain muddling in the character traits of the two cats, wherein Cleo is sometimes referred to as anxious like Jonah, and Jonah is referred to as flighty like Cleo. Such a conflation of the two cats even occurs during a French television broadcast, following the success of *Cleo*, where the reporters mistakenly discuss Jonah as the subject of their broadcast![3]

In spite of all this and the fact that Helen casts herself in the main role in all the humano-feline interactions, based on a certain reality and the fact she is the one narrating the story, reducing the importance played by other humans, the narrative is pertinent for revealing certain ways of being. One must nevertheless refer back to other sources so as to allow for a more cohesive and coherent documentation of our enquiry over a period of three centuries, stopping here in 2012.

— . . . AgiTaTioNS ScREamS SCENTS EnVelOpInG; other similar cats STUCK below; ClAwiNg PuLling PUshINg clIMBinG; STOpping BECoMing StiLL; a DARK MASS in front; Pupil-SPrEadIng Ear-TuRN quIvEr-SnIfF: forms exhalations reassuring vibrations; s-t-r-et-ch-in-g-out-h-is-leg-s s.l.o.w.l.y squeaakkingg; feeling contact regains confidence meeeoowwws; loses his grip; pUShEs, again CLimBS again; LEaPs CLAWs OUT LEgS ABovE head below; withdraws claws, falls pulls . . . push . . . jUMpS at the approach of part of a human with a recognized odor; HiDes, prepares to jum . . . p—suddenly feels himself being captured, picked up, displaced by the odors emanating from above, then placed on a mass that he detects as warm and happily vibrating, curling up inside, purring and glancing at Helen.

Exhausted by the operation to remove her cancerous tumors and having no desire to relive another "motherhood" and subsequent mourning with a new cat, like many cat owners beginning in the last third of the twentieth century, she doesn't want another cat as a replacement, even less a cat coming from an animal shelter. However, she agrees to go visit one, upon insistence of her sister and out of pure curiosity and a need for a distraction, and is quickly fascinated and captivated by this acrobatic cat, climbing up the common cage, and then falling down from the highest rung, sticking out a paw toward her, a cat *she finds beautiful* like a Siamese, thin, slender, covered in a wooly fur, except for her brown face, paws, and tail, and who she believes calls out to her, to which she replies by touching his paw protruding through the mesh of the cage, thus transforming their bodily techniques into a voluntary—and not natural—form of interspecies communication. Helen had already learned to solicit cats (she and her parents had first become sensitive to dogs), and the feline had learned to solicit humans.

Perhaps for the cat, it was more about playing and interacting, wanting to rub against her rather than to be simply taken, but his manners *with her and then her eldest daughter*, rolling around on his back in her arms, convinced them to buy him, therefore granting Jonah's wish as well, placing him in *the principal role in this decision*. More than likely, this had to do with his age, which was greater than that of the rest of his peers in the cage, meaning he already had at least four months to master his locomotion and balance through his acrobatic feats. Perhaps he had acquired such knowledge elsewhere, since the official explanation of conjunctivitis, which led him to be placed in quarantine for several weeks, is later called into question by outside opinions regarding the cat's hyperactivity. *He had perhaps arrived at the shelter by way of an initial family who had returned him in exchange for another cat, but not without partaking in interactions with him.* And his hyperactivity was the result of his initial phase of being cared for at *a kitten farm or cat mill, where he could be sold for a much lower price without the proper paperwork, in an animal shelter not concerned so much with the animal's living conditions.*[4]

> Such cat mills or kitten farms began to appear at the end of the nineteenth century and through the interwar period, most notably in the United States in order to sell luxury cats (Angoras, Persians, or Siamese as is the case here) or to make feline "fantasy" outfits and covers for automobiles. Following a similar trajectory as those institutions dedicated to dogs, these farms saw a large increase in the second half of the twentieth century due to the widespread popularity of cats as pets or as a companion animal. And such an industry had significant consequences for male kittens, often separated too early from their mothers, at the age of two weeks. The mother, having been more or less well fed throughout the pregnancy, would quickly be swept up again to get pregnant once more and therefore hardly showed any concern about the welfare of her new litter, who then

> hardly had any time at all to acquire any skills from her and therefore had to learn through trial and error, a practice that is more stressful and less efficient, leading to behavioral problems: the cats are now either crafty, playful, and interactive with humans but perhaps too hyperactive, as is the case here with Jonah, or they become irritable and aggressive if male and hardly interactive at all if female. Each of these tendencies may then serve as a reason for the cat either being abandoned or receiving clinical treatment, particularly for hyperactive cats whose numbers continue to increase, similar to dogs and . . . children, since they all live within the same Western environmental conditions. If veterinarians of yesteryear hardly ever treated such cases, and if the contemporary veterinarians have become more and more sensitive to identifying such cases in animals—to such an extent that animal psychiatry is in the midst of a boom—it's because the doctors apprehend such disturbances as individual facts. This is partly correct, but they often forget the consequences on the cat population as a whole since such behaviors are not merely found in pets but also spread into the stray cat population. The phenomenon of abandoning cats out in nature as a result of such behavioral issues follows a similar pattern among female cats, without much education, who do the same thing with their male young. And this is one of the reasons for the development of cat interactions with humans . . . the latter also having their role to play in all of this.[5]

Helen embodies this role well in naming the cat Jonah. The cat's athletic grace reminds her of the rugby player Jonah Lomu, worshipped by her husband. This name thus allows for the cat's integration into the family by way of a more common human name—a practice that was once rare, often making use of diminutives such as Pierrot or human nicknames such as Miton. And such a naming practice continued to become more common by the end of the twentieth century, with the increased use of diminutives like that of Lilly and Bob. Helen immediately buys an arsenal of toys for interacting (balls, plastic mice, a fishing pole dangling a fake bird), while simultaneously exclaiming her surprise at such necessities, having never done so with her first cat, indicating that she is giving in to the demands of the market (but the market proposes such toys because the public demands them, convincing themselves to buy them) and realizing that she has changed over the past twenty-five years, since first welcoming Cleo in 1983.[6]

— . . . suddenly feels himself being enclosed and isolated [cage]; pupil-spreading the striated shadows [the mesh]; quiver-sniff the stench surrounding him [plastic]; an agitated ground gripping, ROUNDING HImSelf; scents human screams enveloping him ear-turn; a swarm of vibrations [car] TURNINGINCIRCLES; the ground is picked up stirred grasping with claws; stops whiinnes; pupillates a bright opening, pushes his paws down, hurTLEs himself outside, and just as quickly detects a different scent, slamming on the breaks of paws; shakes off the stench [of plastic] quiver-sniff, de-pupillates, ear-turns; unknown; l-a-u-n-c-h-es his legs toward a dark fold [underneath the recliner] FLATtens Himself out; quiver-sniffs, pupillates, ear-turns; perceives, smell, vibrates; records, measures, adjusts, becomes calm; concentrates: pupillates movements [*legs coming and going, an agitated fake bird*], pounces on it with his paws, and repeats; c-a-st-s his paws out again, grabs hold of it [*the bird*], pounces, clasping in paws; suddenly de-pupillating, quiver-sniffing, a liquid aroma [*a bowl of milk placed before him*] pawing it, quiver-sniffing this paw, shakes his head, pushing it with one of his paws; de-pupillates a human approaching, l-a-un-c-h-e-s his feet afar, clasping, clawing, pulling *[a curtain*]; ear-turns a call, slams on the breaks, stops his paws, de-pupillates the arriving call, l-a-un-c-h-e-s his paws toward a high surface [the top of a cabinet] then toward another spot [*kitchen counter*] . . . —

Already a big cat, standing up on his slender hind legs, Jonah climbs and tears down the stairs or galops through the hallway, surveying and subsequently locating the space, which he perceives as a vast space partitioned into various forms of strong luminosity (*large, bright stained-glass windows: white, apricot, blue, and red*), gathering the humans together below (the living room) and dispersing them into various other rooms on the top floor (the bedrooms); humans with simultaneously different (one male out of four humans) but similar scents, each composed (parents-children, between sisters) along with their different forms and the similar yet different noises they make, allowing him to distinguish between them while still feeling a sense of community. Initially these specific places around the house are attractive to him because he can paw at things and various movements, rip things up on the floor, or high above (the curtains and carpet), but he is also drawn to the presence of flowing water and to the repeated sound it makes [in the toilet for instance], to shiny lights (buckles on shoes), and other things that simply appear (pages from the printer) or move (the computer keyboard). To such a degree that Helen, still worn out by her cancer treatment, intends to return him back to the shelter. Helen had not previously experienced such hyperactivity with her former cat, Cleo, and when adopting Jonah, she had simply wanted the same peaceful and quiet inside cat breed from days gone by.[7]

The suggestions solicited by Helen in light of this issue put the blame squarely on this specific breed of cat, and on Asian cat breeds in general, an idea often reiterated within veterinarian literature. Such a biological explanation—considering cat breeds as natural invariants—calls for a much more nuanced reflection. In the nineteenth century, treatises on cats make no mention of such incorrigible Asian cat breeds. The best-known and widespread cat breed of the time was the Angora, who were deemed to be intelligent but lazy like "all the beautiful breeds," since they were conditioned into being indoor cats. In the interwar period, new Asian cat breeds—first and foremost the Siamese cat—already quite prevalent in Great Britain and France—are also described as intelligent, but they are also independent, hunters, and thieves. And yet, contemporary portraits place a greater emphasis on proximity, solicitation, play, and hyperactivity. Such a change in description does not necessarily mean that veterinarians in the past had poorly understood a breed's demeanor. It simply shows that cats have recently undergone a change in behavior. Judged to be intelligent individuals and therefore already containing a strong potential for interacting with humans, these cat breeds saw a much greater focus turned toward them, albeit not exclusively, thanks to the work of human selection, determined as they were to respond to the requests for such cats. The public at large quickly justified such a demand in claiming that these Asian cat breeds were closer to humans and therefore increased the demand for them. In order to meet the high demand, industrial breeding began to meddle in the amplification of such demeanors and sought to reinforce these tendencies, leading to the problems mentioned above.[8]

And now I think it's a good time to reflect with you about the various factors leading to the spread of such and such types of cat. What caused anthroposized felines to become more widespread and led them to change their behavior given that most cats are often isolated in families of humans, as was the case with inside cats of the bourgeoisie during the nineteenth century? We have seen both the human and animal pressures and processes at the origin of such phenomena. Let's reflect a bit more on the transmission of such behavioral changes. It is not unfounded to think that such changes are not purely biological but perhaps more complex, at the sociocultural level. This would imply that these demeanors were socially transmitted constructions, that is: transmitted cultures.

Obviously, there is a cultural transmission through human individual and social selection: breeders, merchants, and the general public who privilege other individuals adapted to the demand that arises from culture. This selection takes place at the moment of reproduction, through the choice of partners, and then continues to be selected for through the male kittens, who are chosen to be sold, or gifting and exchanging them with others, and eventually even through adoptions in animal shelters. And now, we have seen a shift in the tendency for humans to seek out very active cats in shelters beginning in the twenty-first century, with the less active kittens bearing a larger risk of being euthanized.[9] The succession of generations and the short- or long-term future of the species therefore must traverse through the crucible of the "good cat" based off any given era's current definition. This assures their diffusion while also orienting, for a time, the history of the species. Nevertheless, this human selection is not enough, since reproduction is not always performed by the selected cats themselves, as evidenced by Moumoutte Blanche or Dewey. Not to mention the current state of spaying and neutering of cats, which limits this channel of cultural diffusion.

In some cases, the environment can also intervene in helping with cultural transmission. For the moment, this aspect is still rejected by ethologists. In effect, they will need a clear distinction to be made between biological behaviors or those provoked by the environment and those behaviors uniquely considered as social and cultural, so as to demonstrate the autonomous existence of the latter and get us to accept the presence of animal cultures. And yet, cultural transmission by way of the environment is already completely accepted in the area of the human sciences, which point out the role played by institutions, media, schools, etc., while also having the opposite tendency in contrast to ethologists, namely ignoring the role played by biology! Here, transmission is assured by the humans of a specific era. They propose similar frameworks and relations based off their common culture, through which these cats construct themselves in a similar way. Within the various types of house cats, or those considered as child-cats and catdogs, where the transmission of behavioral traits cannot be assured by the group, the environment serves as an intermediary. It allows for the diffusion

of specific ways of being within space, within similar human families, and within time, in each family, from one generation to the next. The environment therefore transmits culture through an analogous adjustment and an analogous construction, as cats brush up against this environment. And the cats change at their own rhythm in relation to this environment.
Furthermore, these felines perhaps transmit cultural changes through their reproduction, with their epigenetics having registered the subtle changes. As I told you, the hypothesis of how cats transmit cultural changes is under debate. However, if we look at our historical cats, it would seem that the epigenetic transmission of cultural changes takes place over several generations in order to explain them. We should also consider an epigenetics that temporarily transforms the newly acquired skill or trait into something innate while slowly modifying this innate trait with a new skill, which would justify transmission and modification. This would seem coherent and would allow for the insertion of the biological in a dialectic with the environment, that is, specifically with the social and the cultural. And in this process, history can either provide us with some grist for the mill or reject it, in relation to the natural sciences. For example, consider the behavioral modification in Angoras, even if there is a problem regarding an exact definition and continuity of the breed, and even if the homogeneity of this cat breed over two centuries is uncertain. The timescale is too short to think at the level of genomic modification. But it would be long enough to suggest some sort of epigenetic transmission. Otherwise, the findings would imply that the environment reshapes each generation, that a reconstruction of the species begins again with each new generation. But we just saw how male kittens abandoned at a very young age already possess a strong attraction for humans. As if they were predisposed, through a temporary acquired trait, before a welcoming environment could even be played out, and that perhaps the maternal example had already been imprinted.
Indeed, we mustn't forget that besides the epigenetic potential of transmission there is also, if they have time, cultural transmission through the education of the young by their mother, their imitation and observation of adult cats, older siblings, etc. And there is also transmission through the emulation of other felines, similar to how Mouton was able "to

educate" Minette in being an inside house cat. All these various factors become conjugated within an ongoing process of the construction of individual cultures. Those cats who are recalcitrant or rebel against such changes along the way are immediately abandoned by cat owners, who are partisans of a specific kind of cat and relation. Those cats who are more apt, however, at modeling the proper desired traits are sought after, proposed to potential owners, and purchased. Their diffusion is assured by way of this human selection, through a facilitated reproduction, and perhaps through a biological transmission of progressive modifications of the expression of genetic heritage, effectuated in response to environmental pressures. Little by little, these cats adapt their behaviors and their numbers grow if the demand for them also continues to increase. But let's turn our attention back to Jonah, who Helen wants to give back to the animal shelter.

Helen's two daughters are against the idea of returning Jonah; they adopt him due to his beauty and intelligence, along with his turbulence. Such a decision confirms the larger role played by children in selecting pets beginning in the 1960s due to the ever-increasing sway children possess in their families and their preference for active cats, which only contributes to the diffusion of this type of feline. The eldest daughter also helps to soften Jonah's demeanor a bit, placing him on her lap, imposing onto him a more serene attitude, with her singing and Jonah purring, learning interactions that he didn't acquire at the kitten farm and in his cage, becoming attuned to his environment, displaying his affection for everyone. He reinforces this inclination after undergoing a very quick neutering.

Once a very rare occurrence, democratized in the second half of the twentieth century, here proposed as an obvious practice by the veterinarian, quickly accepted by the family, the sterilization of cats became a mass phenomenon in the West by the beginning of the twentieth century, in particular due to the practices of institutions for animal protection, by shelters, and veterinarians all concerned with the excess cat population and the eventual abandonment of the animals. In the early 2000s, the rate of neutering was as high as 80 percent in Great Britain and the United States, dropping to 40 percent in Italy. This tendency is only amplified due to much easier and less harmful ways of performing the procedure, with less risk of unwanted ills or side effects than past operations. Experienced as an individual and biological event for the families and the veterinarian, such a process of pet sterilization has also become a social event for both the humans and the animals. One could think of it as a break with nature. One must above all see it as a sociohistorical event that can be added onto other factors guiding their ways of being, their feline cultures, and which reorients the history of cats by favoring a certain tendency. In effect, such a growing practice leads to dissuading cats from their peers and redirects them toward their humans, encouraging cats that are closer to their owners, asking for their attention more, and behaving more interactively.[10] And this tendency is amplified by another aspect as well, developed by Anglo-Saxon countries of the South Pacific.

From his very first excursions, Jonah's attitude—a vagabond climber and aggressive toward dogs—*pose a problem for Helen's family, which was not the case with their previous cat, Cleo, especially since both the place and the specific feelings attached to their new cat had changed. Helen is now afraid of losing him either due to Jonah fleeing, an accident, or theft in an active city like Melbourne. This cat was supposed to serve as a good companion and is therefore now a very precious thing. Once a minor fear and confined to cats who lived in the bourgeois apartments of big cities, the issue of having one's cat stolen has quickly spread with the increasing value of cats and their social diffusion now including suburban neighborhoods, which for a long time served as the very oasis of cats. Pressure from neighbors who alert cat owners when their cats stray, who provide suggestions, and who demonstrate their approval only reinforces this tendency following a similar sentiment regarding another aspect of cats—when they become a nuisance: Jonah reveals himself to be "an extraordinarily good bird hunter." He ravages the fauna.*

The cat's tendency to destroy fauna is not without importance, since there is a reciprocal and profound transformation in cats as a result of this activity. And such a change has originated over the past several decades in New Zealand to the point that it has become an ecological crisis, and in Australia, which has already had to confront the climate crisis with droughts and degradation to its biodiversity, the two countries converging in their attempts to preserve such biodiversity to the best of their abilities. Through the influence of naturalists, both countries' populations have become more sensitive to important feline predations. Such predations are real and cannot be denied,[11] but their proportions have less to do with the cats themselves (their rate of individual destruction does not appear to have increased) but to their recent proliferation as a result of their success, of being adopted in mass . . . and are a result of their being equally abandoned in large numbers due to poor human management (or lack thereof). In response to this problem, there has been a call to control the feline population through an organized sterilization program, trying to neuter or spay a maximum of cats, to eradicate strays and not simply stray cats in the cities—as was once typical—but also those strays found in rural and suburban areas. There have even been calls for cat owners to keep their cats inside and to put them on a leash once they do go outside with them. To such an extent that the number of cats in Australia has already begun to decrease and 80 percent of cat owners make their cats come inside at night, and 41 percent do so during the day, starting in the 2010s.[12] The program and its application were exported to the United States, where the rate of keeping cats indoor could reach 50 percent,[13] but it has also led to a number of social debates only now beginning to take place in Europe, where the discussion at this point is still merely scientific. However, certain countries (such as Belgium) have already begun to take action.[14] More than likely such measures will be taken following a more serious sensitivity to ecological questions even if a good many cat aficionados in the United States and Europe do not want to recognize cats' tendencies for predation.[15] They consider it as normal for an animal thought to be naturally wild, independent, free, and impossible to keep indoors—the typical portrait infused with the Romantics' sentiment taken for reality whereas the cat *is* according to its environment. This has created (for the moment?) fairly significant distortions in the West: whereas American veterinarians have officially called for the placing indoors of all cats since 2007, most of the European cohorts still consider such measures as a privation, even as an instance of cruelty.[16]

In fact, cats are beginning to experience what dogs in Western countries experienced between the middle of the nineteenth and twentieth centuries, with the eradication of strays—the majority of which for a long time had been dogs—and the enclosure of dogs by their owners, which up until then had only been true for a small minority, a practice that quickly led to the disappearance of freely wandering dogs and transformed the rest into pets. If this same tendency is followed with cats, they will become just like any other dog! They will lose their advantages and . . . their inconveniences . . . and will eventually acquire others!

As such, Jonah quickly experiences difficulties connecting with outside scents, sounds, and vibrations: he collides with obstacles even if the air emanating from them is clearly coming from elsewhere (*colliding with the screens of open windows*) and in some places can only see but not smell (*the closed French doors*); this leads him to having to trace out passages but he quickly finds himself caught and brought back inside. Or he feels his hind legs, back, and chest *snared within a harness equipped with a leash, which Helen purchases at the animal shelter on the advice of merchants adapted to the situation in Australia and which becomes a personal cause for Helen wherein she gets behind the idea of "walking him like a little show dog," whereas she would have never thought of doing this previously with Cleo, who was much more of a flighty cat, on whom she only imposed a collar*. Jonah resists the leash for a long time (several months)— . . . moving his paws at the side in order to "loosen it" . . . rounding his back, trying to make himself smaller, pushing his paws on his hind legs, twisting his body and head in order to free himself on some occasions . . . more often: pull on his paws to quickly connect with other scents and desired movements quickly changing directions and locations . . . "untying himself" through quick turns of direction . . . raising his legs up and turning in circles . . . getting tangled up around the human by jumping from one side to another . . . pushing and pulling . . . —

> Encouraged by various associations of veterinarians and the authors of manuals from the late twentieth century, the practice of using a leash with cats develops in Anglo-Saxon populations, even if it is still a minor practice, like using a collar. It has recently been introduced into the urban centers of Western Europe, diffused and legitimized by the internet. For the reader who is persuaded this is an aberration against nature, fine for dogs and their owners but not for cats—who are free creatures within the depths of their being—we refer back to Balzac who, already in 1839, during an era when dogs were also free to roam, creates the narrative scene of the knight of Halga, mocked by the population of Guérande for being "ridiculous with his dog" because he walks with her on a leash while chasing away the male canine suitors with a cane. As was the case with dogs across several generations, little by little the cats in question learn, understand, and even cooperate.

"I never thought we'd end up with a cat crazy enough to want to go for walks," *writes Helen when Jonah turns four years old*. For he associated the object with going outside and even learns to signify his desire by snaring the leash between his teeth and bringing it to his human. *If the latter ignores him*, "patting my abdomen with his front paw," "he drops the leash at my feet and emits a baleful meow. Irresistible." Outside, he pushes his paws to express his impatience but eventually adapts: dragging her along (*the sidewalks*), stopping here and there, quiver-sniffing, ear-turning, taking off again, having acquired a certain way of being, constructing a culture.[17]

> Having arrived at the end of the present book, dear readers, I'd like to specify once again what I mean by the term culture. I can hear a number of you grumbling from the very beginning, claiming that I haven't explained this term enough. I intentionally only mentioned a precise definition in just two quick sentences at the very beginning, in order for you to slowly grasp its depth throughout the following pages and examples, in a more assured manner than if I were to simply have hammered it home at the beginning. My definition will appear clearer here thanks to memories of the more concrete aspects discussed throughout the pages you've just read.
> First, let's get rid of anthropocentrism. That would not be the case if we were to define culture according to humans in order to then certify that it only applies to them. We must also avoid anthropomorphism. Here again, it's necessary to de-anthroposize the concept, to show to what extent it's a generality, an abstraction. In that light, I have chosen to retain several theories among the different levels of possible meanings, approaches, and observations posited by anthropologists over the past two centuries that have eventually been deemed as flawed. But in making use of them, I have recognized how these anthropological viewpoints must be understood as adaptable and more specifically adapted to animals, as zoo-anthropologists have recently demonstrated. As such, in its most generalized definition, culture is an ensemble of skills that are slowly but surely constructed and learned, but also shared and transmitted. For cats, when we talk about culture, we're not talking about a musical heritage, a scientific heritage, or a technical cultural heritage. Rather, when we speak of cat culture and its heritage, we must understand it as a "know-how-to-be," of the transmission of various ways of being, of behaviors, that are progressively acquired and constructed. This extension of culture to the body, to gestures, to the psychological— as culturalized elements—is typical in the human sciences: from Marcel Mauss's techniques of the body to Erving Goffman and the cultural history of emotions or the gestures and bodies referenced by Alain Corbin.[18]

There is no reason whatsoever to not include animals in such an extension of culture to the body, in this case, cats. This works as long as we don't make them out to be mere biological machines, and if, instead, we focus on their behavioral plasticity within space and time, similar to humans, and admit there are other factors and dimensions at play, as ethologists are slowly beginning to recognize.

You perhaps have noticed that I've chosen to approach the feline not at the scale of the species or group but at the level of individuals. In order to do this, I leaned on the anthropological school of personalism. This school in anthropology considers individual humans as culture, and that culture does not exist outside of them. We must therefore grasp how these individuals live culture, how culture sets them into action, how they then construct themselves and their behaviors, their personalities and cultural types. Such an approach leads us to understand culture not as a whole or a fixed system, but as a dynamic process. And to reflect on this, I also relied on historical anthropology, only retaining the ideas concerning adaption to a given environment of contacts, exchanges, and evolution as well as the anthropology of interactionism, which emphasizes the role played by relations, the various contexts leading one to modify behavior, on transculturations.

I therefore conceive of feline culture as an ongoing process of selection and construction. I have emphasized the process of education, the shaping, and enrichment of this culture, of ways of being, living, and externalizing the culture, and even internalizing it. I focused on the relations with the environment, on the processes of borrowing, adjusting and adapting as well as transmission. All the while I postulated that these ways of being, these constructed cultures are phenomena tied together at several different biological, psychological, cultural, social, and historical levels. I posit such a conception in contrast to ethologists, who tend to consider everything from the viewpoint of biology or anthropologists who tend to consider everything from the viewpoint of culture and others who tend to separate biological behaviors from cultural behaviors.[19] I therefore speak of individual culture when it's a question of constructing individual attitudes under the entangled influence of both the biological and the environment. Such cultures of individuals eventually become group cultures if there is a transmission either by way of animals or the environment. Let's return to Jonah.

His new culture is not merely limited to the pleasure he receives from being walked on a leash, even if this is perhaps the most symbolic part of it. Rather, this new culture is embodied through multiple aspects. *Concerned about a compromise between the feline desire to go outside and the need to control and watch over him, Helen, provoked by suggestions, recent examples, and commercial advertisements, decides to install a cat run in her backyard.* Jonah learns quickly— . . . pushing his head and paws through [*the cat door*] . . . pushes his paws to quiver-sniff fresh scents ear-turning toward swirling noises de-pupillating the trajectory of his movements . . . clawing to his heart's content [*on a wire mesh tower*] . . . allowing him to get closer [*through a tunnel several meters in length*] . . . and smell and obverse at the end of it [*upon arriving to a second, even more substantial tower*] . . . casting out and launching his paws onto another fixed hard elevated surfaces [*ledges*] making supple movements [*across hammocks*] . . . quiver-sniff, ear-turns, pupillates . . . falls asleep . . . sets about again, falls asleep . . . —

Actually, Jonah is more frustrated than he is content. If he is now able to better observe from above—which cats love to do—his desire for getting outside and running off is all the more reinforced as well as his sense of privation when he glimpses prey appearing close by, *which are both completely counter to Helen's objectives.* However, slowly but surely, Jonah changes his focus and eventually seeks out more interactions with humans: "After half an hour lying in the hammock and being tortured by fat pigeons on the fence, he'd trot inside and demand to be stroked or carried."[20]

In reality, from their very first contact, Jonah and his humans learn to interact more so by way of play. Upon his arrival, he will initially only leave his place beneath the bamboo chair in order to chase down and capture a fake bird. Then, when he climbs, jumps, galops, and flees, he will only respond to this interaction of play with the toy bird on a fishing pole. He quickly grasps the need for a human and just as quickly solicits her attention: grabbing the fishing pole in his teeth, pushing his paws in order to hop up *onto the bed where Helen is resting* and "releases it" into her hand. Did he learn this somewhere else or had he watched and quickly understood? *This is a major shift from Helen's previous experience with cats. Not only does Helen play with the cat but she also describes this play in several different ways, aware, through conviction and influence, of the importance of a hybrid socialization, of getting to know each other better and living together. The parallel with children is rather obvious, to the extent that Jonah is already integrated into this symbolic order, the adults behaving as his relatives, and her daughter, as mother and sisters.*

As such, feline and humans quickly learn to work together. When Jonah grabs the bird with his teeth or paws at it, he just as quickly releases it in order to push his paws again: grasping it, letting it go, as he would do with a prey animal during the initial education of male kittens learning how hunt from their mothers. A task he more than likely never learned, but which he replaces by way of cognitive dispositions adapted for running off any agitation and from which he receives pleasure. *When Helen becomes tired* and stops playing, he grabs the fishing pole by his teeth and not the bird, releasing the pole into Helen's *hand*, soliciting her attention in the most concrete way, the most inter-understood. The second day, he purrs as soon as he releases it into this same *hand*—the same hand upon which he pushes his paw pads and not his claws—waiting for her to release the pole in turn, sending to her two intentionally friendly signals, speaking through the gestural, *which Helen understands quite well.* When the agitation stops too early, he quickly adopts a "strange sound through his nose—a cross between a cluck and a sneeze," *which the family eventually deciphers* as a signal of deception.

Very quickly, he begins to prefer another human—the male—who reacts more quickly, making the chase more difficult, the pleasure more intense. Jonah no longer solicits *Helen* for other games except when the man is absent. This activity leads Jonah to discriminate and choose, helping him to differentiate and singularize, perhaps just as much as he does with scents or forms. It also invites him to socialize with his humans. A fundamental task for kittens (and children) in order to grasp the world and integrate its norms, the game allows him, *as it does the family*, to observe the other, to draw a parallel between each of their gestures, to decipher intentions, cooperate, and to learn better and more quickly.

Jonah experiences this with another game he'll learn later on, consisting of a soft texture—*a florist ribbon, always from the same specific flower shop*, no doubt possessing a shiny coating that attracts him or a particular euphoria-inducing scent. He grabs it in his teeth, then releases it into a hand to solicit game play, to then run after the stirrings it makes on the ground and in the air between pillows. *Helen understands this feline pleasure and buys the one type of acceptable ribbon from the one florist who uses it, recognizing the role of play in the creation of a hybrid community comprising two species. In order to maintain and develop this community, to interact, and counter the "guilt and failure of Jonah being an indoor cat," to entertain him, Helen purchased ping-pong balls for him to pat and chase through the maze, and other balls with bells, big balls with batteries making them roll around mysteriously, fake mice dipped in catnip . . . all of which she had never done for Cleo—a more independent and less integrated cat. Although loved all the same. The community of games is therefore dynamic, endlessly renewed and enriched, such as the example of the "nurse" who teaches Jonah* how to play hide-and-seek for cat treats hidden inside rolls of toilet paper.[21]

Kept inside and entertained, Jonah begins to follow *Helen around, who is always present*. He also seeks out physical contact with her, enjoying laying on her belly during naptime, each of them exchanging purrs and *vibrations*, sleeping on *her lap in the morning while she writes at her desk, cuddling next to her while she watches television on the couch*, carefully observing the movement of images on the television screen along with her . . . above all when what they are watching contains dogs! Later on, having become dependent and disturbed by her absence, he follows her while meowing until *Helen decides to place a cloth supermarket bag over her shoulder and put him inside, letting Jonah travel with her wherever she goes*. "Cocooned in the bag, he stopped meowing and started purring. The rhythm of my footsteps soothed him. With his head peering out of the top of the bag, he saw everything that was going on and was comforted."[22]

Helen consents to this, even initiates it, at once surprised at being "ruled by a cat," which hadn't been the case with Cleo, who was much different (as was Helen during that time period), and delighted by this contact that also pleases her, these interactions with a cat she views as intelligent because he begs, thus sharing in this recent desire of felines. Helen demonstrates the role played by women in the contemporary context, emphasized in a number of studies, who pay attention, decipher, and take into account their cats' personalities, taking the initiative to know them, even having more of a relationship with cats than men. As our research has already noted, we have seen that such a feminine role is not new, but it appears to have gained trac-

tion more recently and become more widespread thanks to the trend of having a cat companion, integrated into the family, no doubt because these women don't hesitate (at least compared to their predecessors) to consider the cat as a child. As such, Helen, who affirms living another maternity with the cat, gets the idea for carry the cat around in a tote bag because that's what she had done with her daughters when they would cry or become upset, thus soothing them through her warmth and proximity, and she simply transposes this technique onto the animal.[23]

Jonah learns to communicate via several different registers. Besides the communication he emits spontaneously, emotional signs *that Helen feels and in part deciphers, gestures and cries, even scents and vibrations, as well as pheromones*, he also adopts others he feels are efficient based on experience—for example, meowing in order to be able to lie down on his human or get carried, rubbing up against someone in order to open the office. Other than the signs (scents, postures, vibrations) that she also emits (like all humans) without noticing, but which cats can read, *Helen adopts other gestures as well.* Most notably, she speaks to her cat. Since this form of communication is not deemed as aberrant and is a more common practice than in the previous eras (similar to dogs and . . . children), her book transcribes some of these phrases. It is highly unlikely that they were the exact words uttered by Helen due to the very lag in the publication timeline, but what it reveals is a certain register, a manner. It's the daily use of the first name, Jonah, that slowly became common during the twentieth century, more important for humans to say in terms of integration than for the cats themselves, where once "cat" or "pussy" would have been enough. It's also a form of a more familiar informal address that allows for a larger integration of the cat into the family, whatever social milieu they might be a part of, used by each member of the family to refer to the cat beginning in the 1960s. The use of the term "boy" or "buddy" also accentuates the integration of the animal, circulating a form of affection, an emotion that the feline is also capable of grasping. Everything must be conveyed by short, simple phrases (commanding, reassuring, asking) through which cats (or dogs or children) are more apt to understand as long as there is intonation. For an animal such as Jonah reacts to words through their diction, to their sound, vibrations, scent, and the various microgestures he perceives given his close proximity to the human within the phylogenetic tree of life. Which doesn't mean that he can't also, through association, understand and memorize specific words, thus adding yet another level of communication.[24]

In fact, Jonah manifests a need for human attention that goes beyond Helen and her family, extending out to others if they are in his territory and well-intentioned. For a while, he becomes attached to some painters, clawing at their brushes and observing their every move. "With their white overalls, stealthy movements, and penchant for climbing, the painters must have seemed like human cats to Jonah," *Helen intelligently notes.*[25] He also tries to get their attention during coffee breaks, meowing softly, climbing up onto the kitchen counter and tapping his paws on their faces. And each time, *he receives love in return.* Jonah also begins to wait for these humans every morning and meows for several days after they have finished their work and gone. Besides the need for attention, Jonah also requires activity, which *his humans appreciate and solicit but to which they don't always respond.* Confined as he is indoors, Jonah can only pretty much engage in activities *with them.* Unlike Dewey, he can't simply replace one *defiant human with another accepting one.* He gives in to other substitute actions (*destroying the carpet or clothes, stealing objects, and swallowing rubber bands . . .*), also a way to signal stress in relation to his environment, like the day when he wreaks havoc in *Helen's study*, counter to which he "moans" when *she calls to him.*[26]

> Jonah embodies various attitudes that are becoming more and more signaled by cats who are forced to remain indoors. In contrast to what many veterinarians and ethologists believe, these new signals are not a result of being confined (they weren't signaled in the nineteenth century by house cats who were much less active and cultivated to be lazy); rather, they are a result of a more recent selection for cats with more active temperaments. Furthermore, studies have shown that the signaled pheromone concerns primarily (for the moment?) cats in Anglo-Saxon countries, precisely where these new cats were bred. Jonah experiences some of the advocated solutions, in particular an enrichment of the environment first created in zoos in the middle of the twentieth century, then transposed onto industrial farming livestock, and more recently onto pets, in this case used to take into account these new temperaments and resolving any resulting issues.[27]

As such, Jonah is capable of playing by himself with moving and appealing objects (automatic toys) as well as high surfaces (potted plants), big enough for him to stretch out on them and to grasp hold of them, to eventually hone his climbing skills, play, mark them, and express his presence, high enough to let him observe human activities for hours (*cooking, dinner, dish washing*), all interesting enough to channel his needs. And this shift in behavior is how he stops climbing on the kitchen counter, ceasing to claw arms and objects, instead seeking out a higher site for observation (*the ceiling plants*), *constantly rearranged in the kitchen so as to satiate this need.* To such an extent that he is now allowed back into the study, which was forbidden for a time, after the debacle where he destroyed some items, and in front of which he would often meow. He was not content merely to receive a series of long *gentle strokes from Helen, who thought she was pleasing him before closing the door. Rather, her actions simply reinforced* his stress and solitude as well as inaction.

For the humans, such inconveniences of an active cat always seeking their attention is overcome due to his demonstrative affection, especially toward Helen's daughters, at the heart of the family, particularly the second daughter, still in school, who Jonah welcomes each evening on her return home and against whom he snuggles, curling up around her neck. Obviously, he also enjoys the material benefits thanks to such a proximity, such as warmth and human vibrations, which are reassuring for him. But the latter of these arise from an inverse relation: from the positive effects he provides to *the human.* This exchange goes in both directions, where each species appreciates the other and demonstrates this—*through speech and petting on the one side*, and through purring and rubbing up against on the other. This attachment even leads to watching over her, for example *when the daughter is ill*, even alerting the parents, through a series of "loud yowls," during an evening crisis where their daughter required more pain medication, *leading Helen to exclaim that it would be impossible to get rid of "part of the family."*[28]

And Jonah has trouble with family separations. He quickly shows himself to be quite sensitive to *daily absences*, waiting at a window if he is alone, running to the front door to welcome the *daughters and Helen's husband in the evening*. Then, he begins to associate quick *back-and-forth movements and suitcases with prolonged absences*. When this occurs, he climbs up the stairs and constantly jumps, meowing and circling around the *person suspected of departing*, lies down in an open bag, or sticks closely to the *baggage brought to the front door*. Conversely, whenever *the person returns*, he doesn't want to leave their side, purring with pleasure, as a sign of friendship, rubbing against their clothes or skin or curling up in a ball on their lap, so as to reconstitute a mixed scent of the community, following them everywhere . . . like a dog! He enters into a stressful state during a long absence by Helen's husband, urinating in various places to indicate his discontent, resisting *the solutions proposed by the veterinarian, such as using a diffuser that would spread reassuring pheromones*. And then *the return*, which he can foresee, no doubt *informed* by a change in Helen's state, wandering around the house, then meowing right before his arrival, more than likely detected at a distance, then reducing but not entirely ending this practice perhaps out of fear of another departure. Anxiety medications stop this disorder, but not his separation anxiety *diagnosed by veterinarians. To such a degree that the family eventually takes to hiding the luggage, filling it up and taking it outside on the sly while another family member distracts Jonah*. No doubt in vain because he could still detect *the restless state of the humans*. He urinates again the night preceding *Helen's departure for her worldwide promotional book tour for* "Cleo."[29]

> Stress and anxiety medication are not merely reserved for catdogs. Such treatments had already been developed in Europe prior to the interest in using them on these specific felines. Such medication is also used in cats who are having problems adjusting to their environment or their relations with humans, such as with stray cats or with inside cats who are independent but who have mediocre living conditions. And this therapy is applied to cats in the same way it was first applied to pet dogs (or animals in a zoo) and initially prescribed for humans—who are rather good consumers of anxiety medication. Such an approach has only increased since veterinarians have discovered states of well-being—as a result of an inner balance (homeostasis)—or unease caused by other factors such a stress in order to find solutions and new forms of treatment. The current development of animal psychiatry has only accelerated readings around the subject and increased those seeking help for their pets. In other words, cats consume anxiolytics because they are stressed *and* because humans have a treatment for stress.[30]
>
> However, we shouldn't believe that the treatment of cats with such medication is leading to a concrete and alarming step toward an ongoing denaturation of the species—proof of human folly in regard to the various states of alley cats, street cats, inside cats, and farm cats from the past whose portraits we sketch with an idyllic view due to nostalgia, habit, and misunderstanding, viewing them as natural and normal. The stress

of urban cats from yesteryear, starving, suffering, and scrawny, constantly hunted, was of a much greater magnitude but certainly real (Camus evokes cats driven by hunger, inspecting garbage with the highest vigilance fearing capture and being brutalized by children),[31] just like that of inside cats, praised by Baudelaire, fleeing their enemy while they were asleep, or that of farm cats and stray cats in the countryside who were hunted just as often, chased after and having very short lifespans, even still to this day. If veterinarians received so-called natural cats for a consultation, they would judge them to be just as stressed, the fear of being trapped being equal to that of relegation!

Nevertheless, the fact remains that the increasing dependence of cats on humans, including the more recent separation anxiety of catdogs, shows to what extent the contemporary cat's connection with humans is vital for both their physical as well as mental well-being. And the same can be said for dogs. As such, this new vital connection between cats and their owners translates a total reversal of ways of being. For it seems cats have been following a path identical to pet dogs but with a temporal lag. Through a progressive construction, through social generations, and anthroposized cultures, the development of cats is not based around humans and no longer a territory. Within such a context, the term "catdog" obviously does not mean that cats are going to become dogs; the very species barrier would prevent such a thing. But they will begin to act like dogs. The term designates a process. And yet, this phenomenon of the catdog is already being introduced into Western Europe, alongside other cat species that have already been anthroposized. This process could indicate the future for cats if the eradication of stray cats were to proceed at the same time. In other words, if we want to reflect on this future, not in looking toward the past in the name of some genetic or physiological connections, we shouldn't name these anthroposized cat *Felis silvestris catus*, as has paradoxically happened, but rather *Felis hominis catus*!

End of the Journey

Throughout the course of this feline adventure, there are several persons who merit an acknowledgement. I would first of all like to thank Svérine Nikel and Patrick Boucheron for welcoming my work into the social sciences division of Éditions du Seuil and more specifically Patrick's illustrious collection, "L'Univers historique." I initially discovered this wonderful collection many years ago as a high school student. I stumbled across *L'Homme devant la mort* by Philippe Ariès, and in my mind, this collection will forever be tied to the beautiful intellectual inventiveness of the 1970s. With a great freedom and open-mindedness, Svérine and Patrick encouraged me and accepted my two strange Oddities into their collection—*Animal Biographies* as well as this book. I hear you lamenting, dear readers: "And what about us? Do we not deserve any thanks for having followed you along on your journey and read your works?" Most certainly, dear readers! A thousand thank yous! It has been a pleasure to partake in this adventure with you—your friendly, interested, patient, and comprehensive nature has been most welcome! What's that you say? You would like to set off again on yet another adventure? I would certainly be happy to join you again! But don't forget the cats: give your cat a gentle caress, the one sitting by your side, who read the book along with you.

> Without predicting the future of felines, the current catdogs, attached to more attentive humans (which doesn't prevent some from being abandoned!), symbolize an inverse situation than that of their ancestors from the eighteenth century, who were hunted by those with bad intentions and who had to constantly flee humans.

For a long time, Jonah sat staring out a window.[1] He carefully observed the path on the horizon by which any new arrivals always pass. He watches and carefully observes. He was probably alerted, informed by postures, bodily vibrations, particular sounds and scents of Helen's husband and younger daughter. Their vibrations and mood alert him to the return of Helen's daughters: to the exact day and the approximate hour of their arrival. Helen sends different signals every day, but similar to the ones for every precious arrival, which he associates with them. Perhaps he pupillated the comings-and-goings, the movement of objects, the scent of cleaning products used at the last minute to prevent any form of criticism. He perhaps ear-turned toward the sounds (first names) attached to the absent humans. Observing the outside listening and feeling the inside. Suddenly head-turn ear-turn pupillating: loud recognize movement approaching (the car); pUSHIng on his paws, RounDing his BAck, rAISIng and MoVinG his tail, raising his HEAD POINTEARS, every sense on alert: pupillating humans, their silhouettes, familiar feelings, perhaps encouraged by *the background agitation*; recognizes a specific form of certain gestures (the eldest daughter who comes down the stairs running) associating it with past scents of warm interactions; rAisEs hEaD FuR TaIl EARs . . . LAuNch Push MoVeS HiS PaWs GAllOpinG toward the site of apparition (the front door) . . . IMmobilizes HIS PAWS . . . ear-turns NOISE QUiVer-SnifF the scent behind it . . . turns-around mee-ooww-ss . . . pushes his head through an opening . . . pupillating fur-vibrating solicitation (arms held out smiles cries emanations of joy) . . . and so he s-o-a-r-s . . .

NOTES

The Cat Massacres of the Eighteenth Century

1. See Marcus Schneck and Jill Caravan, *Le Chat* (Paris: Solar, 1991), 28; Dennis C. Turner and Patrick Bateson, eds., *The Domestic Cat*, 3rd ed. (Cambridge: Cambridge University Press, 2014), 39–40, 52–53 (except when otherwise indicated, the references cited in the book refer to this third edition).

2. Nicolas Contat, *Anecdotes typographiques* (1762), in Philippe Minard, *Typographes des Lumières* (Seyssel: Champ Vallon, 1989), 234.

3. Turner and Bateson, *Domestic Cat*, 197.

4. Contat, *Anecdotes typographiques*, 224–233.

5. François-Augustin de Moncrif, *Moncrif's Cats*, trans. Reginald Bretnor (New York: A. S. Barnes, 1965); Jean-M.-M. Rédarès, *Traité raisonné sur l'éducation du chat domestique* (Paris: Bourayne, 1835), 57–59; Alexandre Landrin, *Le Chat* (Paris: Carré, 1894), 31.

6. Contat, *Anecdotes typographiques*, 233–234.

7. Ibid., 234–235.

8. Robert Darnton, *The Great Cat Massacre and Other Episodes in French Cultural History* (New York: Basic Books, 2009), 88–120.

9. Moncrif, *Moncrif's Cats*; Laurence Bobis, *Une Histoire du chat* (Paris: Seuil, 2006).

10. Contat, *Anecdotes typographiques*, 233–234.

11. Bobis, *Une Histoire du chat*, 189–239. Concerning the Western character of this liaison see Katharine Rogers, *The Cat and the Human Imagination* (Ann Arbor: Michigan University Press, 1998); Clemens Wischermann, ed., *Von Katzen und Menschen* (Konstanz: UVK, 2007).

12. Georges-Louis Leclerc de Buffon, "The Cat," *Buffon's Natural History*, vol. 6 (London, 1797), 2–4, available from Internet Archive, https://archive.org/details/b28775892_0006/page/2/mode/2up?q=cat.

13. Ibid., 6.

14. Ibid., 6.

15. Jean de La Bruyère, *Les Caractères*, xii, 21 (1690; Paris: Le Livre de Poche, 1995), 460; Moncrif, *Moncrif's Cats*, 19; James Boswell, *The Life of Samuel L. Johnson* (Chicago: Scott, Foresman, 1923), 470.

16. Moncrif, *Moncrif's Cats*, 21–22; Contat, *Anecdotes typographiques*, 233; Bobis, *Une Histoire du chat*, 235.

17. See Turner and Bateson, *Domestic Cat*, 52, 207–208; Marta Armat, Tomàs Camps, and Xavier Manteca, "Stress in Owned Cats: Behavioral Changes and Welfare Implications," *Journal of Feline Medicine and Surgery* 18, no. 8 (2016): 577–586.

18. Jean François and Nicolas Tabouillet, *Histoire de Metz* (Nancy: Lamort, s.d. [1775]), 3:187 (quotes 1–2, 6–9); Jean François, "Dissertation sur l'ancien usage des feux de la Saint-Jean et d'y brûler des chat à Metz" (manuscript, 1758), *Cahiers Élie Fleur* 11 (1995): 49–69 (quotes 3–5); *Chronique des boulangers*, 1661–1662, cited by Marie-Claire Mangin, "Le Sacrifice des chat messins," *Cahiers Élie Fleur* 11 (1995): 87.

19. Bobis, *Une Histoire du chat*, 249–255; Wischermann, *Von Katzen und Menschen*, 33–88; James A. Serpell, "Domestication and History of the Cat," in Turner and Bateson, *Domestic Cat*, 94–99.

20. Francois and Tabouillot, *Histoire de Metz*, 187; Mangin, "Le Sacrifice des chats messins," 97–102.

21. Moncrif, *Moncrif's Cats*, 99, 111–115, 119–127.

22. Sir Keith Thomas, *Man and the Natural World: Changing Attitudes in England 1500–1800* (New York: Oxford University Press, 1996). See the chapter "Man and Animals."

23. Moncrif, *Moncrif's Cats*, 70, 101.

24. Turner and Bateson, *Domestic Cat*, 72, 105–106; K. J. Barry, "Gender Differences in the Social Behavior of the Neutered Indoor-Only Domestic Cat," *Applied Animal Behavioral Science* 64 (1999): 193–211; Filip Jaros, "Cats and Human Societies," *Biosemiotics* 9, no. 2 (2016): 287–306.

25. Buffon, *Buffon's Natural History*, 6:6.

Part 1. A Withdrawal into Territorial Cultures

1. Olof Liberg, M. Sandell, Dominique Pontier, and Eugenia Natoli, "Density, Spatial Organization and Reproductive Tactics in the Domestic Cat and Other Felids," in *The Domestic Cat*, ed. Dennis C. Turner and Patrick Bateson (Cambridge: Cambridge University Press, 2014), 119–147.

2. Marie-Amélie Forin-Wiart, "Identification des facteurs de variation et de la prédation exercée par les chat domestiques (Felis silvstris catus) en milieu rural" (thesis in biology, Université de Reims Champagne-Ardenne, 2014).

Chapter 1. Street Cats

1. Janet M. and Stever F. Alger, *Cat Culture: The Social World of a Cat Shelter* (Philadelphia: Temple University Press, 2003).

2. Paul Léautaud, *Bestiaire* (1959; Paris: Grasset, 1990). There are, however, some brief references here and there in his published journal and not in his bestiary, but they are not indexed. See Paul Léautaud, *Journal littéraire* (Paris: Mercure de France, 1986), 3 vols., and *Journal littéraire: Histoire du journal, pages retrouvées, index general* (Paris: Mercure de France, 1986). Regarding writings on Léautaud, see Marie Dormoy's preface for Léautaud's *Bestiaire*, 7–39; Martine Sagaert, *Paul Léautaud* (Paris: Castor Astral, 2006), 70, 72, 101, 112, 114.

Wandering, Independent

1. Paul Léautaud, *Bestiaire* (Paris: Grasset, 1990), 41, 43.

2. Marcus Schneck and Jill Caravan, *Le Chat* (Paris: Solar, 1991), 40; Dennis C. Turner

and Patrick Bateson, eds., *The Domestic Cat* (Cambridge: Cambridge University Press, 2014), 39–40.

3. Turner and Bateson, *Domestic Cat*, 41–46; *Élever et soigner son chat* (Paris: Larousse, 2015), 22.

4. Marie-Amélie Forin-Wiart, "Identification des facteurs de variation et de la prédation exercée par les chat domestiques (Felis silvstris catus) en milieu rural" (thesis in biology, Université de Reims Champagne-Ardenne, 2014), 3–4, 19–21; Sarah Bortolamiol, Richard Raymond, and Laurent Simon, "Territoires des humaines et territoires des animaux," *Annales de géographie* 716 (2017): 387–407; Pierre Bonte and Michel Izard, eds., *Dictionnaire de éthologie et de l'anthropologie* (Paris: PUF, 2007), 704–705.

5. Elisabeth M. Mesters, Philip J. Seddon, and Yolanda van Heezik, "Cat Exclusion Zones in Rural and Urban-Fringe Landscapes," *Wildlife Research* 37, no. 1 (2010): 47–56; Jeff A. Horn, Nohra Mateus-Pinilla, Richard E. Warner, and Edward J. Heske, "Home Range, Habitat Use, and Activity Patterns of Free-Roaming Domestic Cats," *Journal of Wildlife Management* 75, no. 5 (2011): 1177–1185; Turner and Bateson, *Domestic Cat*, 64–66.

6. Penny L. Bernstein and Mickie Strack, "A Game of Cat and House: Spatial Patterns and Behavior of 14 Domestic Casts (*Felis catus*) in the Home," *Anthrozoös* 9, no. 1 (1996): 25–39; Janet M. and Steven F. Alger, "Cat Culture, Human Culture," *Society & Animals* 7, no. 3 (1999): 199–218; Margaret Slater, Laurie Garrison, Katherine Miller, Emily Weiss, Kathleen Makolinski, Natasha Drain, and Alex Mirontshuk, "Practical Physical and Behavioral Measures to Assess the Socialization Spectrum of Cats," *Animals* 3, no. 4 (2013): 1162–1193; Turner and Bateson, *Domestic Cat*, 66–67.

7. Léautaud, *Bestiaire*, 51, 121, 197–198 (quote 2), and *Journal littéraire: Histoire du journal* (Paris: Mercure de France, 1986), 77 (quotes 1, 3); Turner and Bateson, *Domestic Cat*, 11–31, 57, 81; Oxána Bánszegi, Peter Szenczi, Andrea Urritia, and Robyn Hudson, "Conflict or Consensus? Synchronous Change in Mother-Young Vocal Communication across Weaning in the Cat," *Animal Behavior* 130 (2017): 233–240.

8. Léautaud, *Bestiaire*, 42, 55, 66, 122–123, 18; Turner and Bateson, *Domestic Cat*, 26, 192–13, 207–209.

9. Léautaud, *Bestiaire*, 43–44, 40. 55, 57, 80–81, 121–122 (quote), 132, 144; Martine Sagaert, *Paul Léautaud* (Paris: Castor Astral, 2006), 102. See Jean-Yves Bory, *La Douleur des bêtes* (Rennes: PUR, 2013).

10. James Joyce, *Dubliners* (New York: Penguin Classics, 1993); Virginia Woolf, *A Room of One's Own* (Boston: Mariner Books, 1989), 13; Virginia Woolf, *The Years* (Oxford: Oxford University Press), 17; Gustave Flaubert, *Bouvard and Pécuchet*, trans. Mark Polizzotti (Bloomington, Ind.: Dalkey Archive Press, 2005), 289; Emile Zola, *Thèrese Raquin*, trans. Robin Buss (New York: Penguin Classics), 180; Marcel Proust, *The Guermantes Way*, trans. Mark Treham (New York: Penguin Books, 2002), 554; André Gide, *If It Die: An Autobiography* (New York: Vintage, 2002), 91; Albert Camus, *The First Man*, trans. David Hapgood (New York: Alfred A. Knopf, 1995), 139–140; Maurice Genevoix, *Trente Mille Jours* (Paris: Seuil, "Points," 1980), 22.

11. Jean-M.-M. Rédarès, *Traité raisonné sur l'éducation du chat domestique* (Paris: Bourayne, 1835), 8, 71, 94–95; Alphonse Toussenel, *L'Esprit des bêtes* (Paris: Librairie sociétaire, 1847), 216–219 (quote); E. Sanfourche, *Les Chiens, les chats, et les oiseaux* (Paris: Auteur, 1866); Georges Docquois, *Bêtes et gens de lettres* (Paris: Flammarion, 1896), 144, 215.

12. Léautaud, *Bestiaire*, 117, 122.

13. Ibid., 41 (quote 1), 42 (quote 3), 43, 51 (quote 2), 55, 59, 65–67, 108, 110 (quote 5), 122, 197, and Léautaud, *Journal littéraire: Histoire du journal*, 77 (quote 4); Alain Corbin, Jean-Jacques Courtine, and Georges Vigarello, eds., *Histoire de virilité*, vol. 2, *La triomphe de la virilité* (Paris: Seuil, 2011).

14. Éric Pierre, *Amour des hommes—amour des bêtes: Discours et pratique protectrices dans la France du XIXe siècle* (thesis in history, Angers, 1998); Éric Pierre, "La zoophilie dans ses rapports à la philanthropie, en France, au XIXe," *Cahiers d'histoire* 42, nos. 3–4 (1997): 655–675; Éric Pierre, "Reformer les relations entre les hommes et les animaux: Function et usages de la loi Grammont en France (1850–1914)," *Déviance et Société* 31, no. 1 (2007): 65–76; Christophe Traïni, *La cause animale (1820–1980)* (Paris: PUF, 2011); Marie-Pierre Horard and Bruno Laurioux, eds., *Pour une Histoire de la viande* (Rennes: PUR, 2017); Toussenel, *L'Esprit des bêtes*, 218; Pierre Loti, *The Story of a Child*, trans. Carolyn F. Smith (Boston: C. C. Birchard, 1901), 130; Céline, *Maudits Soupirs pour une autre fois* (Paris: Gallimard, 2007), 148.

15. Léautaud, *Bestiaire*, 60, 89, 117 (quote).

16. Claudia Mertens and Dennis C. Turner, "Experimental Analysis of Human-Cat Interactions during First Encounters," *Anthrozoös* 2, no. 3 (1988): 83–97; Dominique Pontier, Pierre Auger, Rafael Bravo de la Parra, and Eva Sánchez, "The Impact of Behavioral Plasticity at Individual Level on Domestica Cat Population Dynamics," *Ecological Modelling* 133, nos. 1–2 (2000): 117–224; Manuela Wedl, Barbara Bauer, Dorothy Gracey, Christine Grabmeyer, Elisabeth Spielauer, Jon Day, and Kurt Kotrschal, "Factors Influencing the Temporal Patterns of Dyadic Behaviours between Domestic Cats and Their Owners," *Behavioral Processes* 86, no. 1 (2011): 58–67; Turner and Bateson, *Domestic Cat*, 115–226.

17. Janet M. and Steven F. Alger, "Beyond Mead: Symbolic Interaction between Humans and Felines," *Society and Animals* 5, no. 1 (1997): 65–81; Nicolas Claidère and Dominique Guillo, "Comment articuler les sciences de la vie et les sciences sociales à propos des relations humaines/animaux?" *L'année sociologique* 66, no. 2 (2016): 385–420; Chloé Mondémé, "Extension de la question de 'l'ordre sociale' aux interaction homme/animaux," *L'année sociologique* 66, no. 2 (2016): 319–350; Eduardo Kohn, *How Forests Think: Toward an Anthropology beyond the Human* (Berkeley: University of California Press, 2013).

18. Alger and Alger, "Beyond Mead"; Alger and Alger, "Cat Culture, Human Culture"; Janet M. and Steven F. Alger, *Cat Culture: The Social World of a Cat Shelter* (Philadelphia: Temple University Press, 2002).

19. Paul Ricoeur, *Time and Narrative*, vols. 1, 2, and 3, trans. Kathleen McLaughlin and David Pellauer (Chicago: University of Chicago Press); Ivan Jablonka, *L'Histoire est une littérature contemporaine* (Paris: Seuil, 2014); Lucile Desblache, ed., *Écrire l'animal aujourd'hui* (Clermont-Ferrand: PU Blaise-Pascal, 2006); Margo DeMello, ed., *Speaking for Animals, Animal Autobiographical Writing* (New York: Routledge, 2013).

20. Elisabeth de Fontenay and Marie-Claire Pasquier, *Traduire le parler des bêtes* (Paris: L'Herne, 2008); Éric Baratay, "Écrire des biographies animals," in *L'Animal et l'homme dans leurs representations*, ed. Sandra Contamina and Fernando Copello (Rennes: PUR, 2018), 163–176; Baratay, "La recherche du point de vue animal," in *Portrait animal*, ed. Sandra Contamina et al. (Rennes: PUR, forthcoming); Baratay, "Penser

les individus," in *Les Études animals sont-elles bonnes à penser?* ed. Aurélie Choné, Isabelle Iribarren, Marie Pelé, Catherine Repussard, and Cédric Sueur (Paris: L'Harmattan, 2020), 55–79.

21. Léautaud, *Bestiaire*, 41, 55 (quote 1), 59, 66 (quote 2), 122, and *Journal littéraire: Histoire du journal*, 77; Sarah Lowe and John W. S. Bradshaw, "Responses of Pet Cats to Being Held by an Unfamiliar Person," *Anthrozoös*, 15, no. 1 (2002): 69–79; Turner and Bateson, *Domestic Cat*, 40–51.

22. Léautaud, *Bestiaire*, 64–71, and *Journal littéraire: Histoire du journal*, 77; Zana Bahlif-Pieren and Dennis C. Turner, "Anthropomorphic Interpretations and Ethological Descriptions of Dog and Cat Behavior by Lay People," *Anthrozoös* 12, no. 4 (1999): 205–210; Olof Liberg, M. Sandell, Dominique Pontier, and Eugenia Natoli, "Density, Spatial Organization and Reproductive Tactics in the Domestic Cat and Other Felids," in Turner and Bateson, *Domestic Cat*, 119–147.

23. Léautaud, *Bestiaire*, 44, 47, 49–56, 61, 67, 72–75, 78, 84, 102, 123, 205 (quote); Frances Simpson, *The Book of the Cat* (London: Cassel, 1903), 32, 34; Éric Baratay, "Chacun jette son chien!" *Romantisme* 153 (2011): 147–162.

Abandoned, Suspicious

1. Paul Léautaud, *Bestiaire* (Paris: Grasset, 1990), 47, 51 (quote 1), 64 (quote 2), 72, 74.

2. Jakob von Uexküll, *A Foray into the Worlds of Animals and Humans*, trans. Joseph O'Neil (Minneapolis: University of Minnesota Press, 2010); Thomas Nagel, "What Is It Like to Be a Bat?" *Philosophical Review* 83, no. 4 (1974): 435–450, who also took on an opposing viewpoint to this article in *The View from Nowhere* (New York: Oxford University Press, 1986).

3. James Gibson, *The Ecological Approach to Visual Perception* (New York: Psychology Press, 2014); Francisco Varela, Evan T. Thompson, and Eleanor Rosch, *Embodied Mind: Cognitive Science and Human Experience* (Cambridge: MIT Press, 2016); Tim Ingold, *Marcher avec les dragons*, trans. Pierre Madelin (Brussels: Zones sensibles, 2013).

4. Camille Chamois, "Les Enjeux épistémologiques de la notion d'Umwelt chez Jacob von Uekül1," *Tétralogiques* 21 (2016): 171–194; Claude Romano, "La Revolution copernicienne de James J. Gibson," afterword to James J. Gibson, *L'Approche écologique de la perception* (Paris: Dehors, 2014); Bernard Darras and Sarah Belkhamsa, "Faire corps avec le monde," *Recherches en communication* 29 (2008): 125–146.

5. *The Epic of Gilgamesh*, trans. Maureen Gallery Kovacs (Stanford, Calif.: Stanford University Press, 1989); *The Song of Roland*, trans. Arthur S. Way (Cambridge, Cambridge University Press, 1913); Anne Simon, "Les Études littéraires française et la question de l'animalité," *Épistémocritique* vol. 13 (2014), online.

6. Léautaud, *Bestiaire*, 46–47, 51 (quote 3), 52, 60 (quote 1), 60–73, 80, 87, 89–90; and Léautaud, *Journal littéraire* (Paris: Mercure de France, 1986), 1:627 (quote 2); Marie-Amélie Forin-Wiart, "Identification des facteurs de variation et de la prédation exercée par les chat domestiques (Felis silvstris catus) en milieu rural" (thesis in biology, Université de Reims Champagne-Ardenne, 2014), 21.

7. Léautaud, *Bestiaire*, 52 (quote 1), 55 (quotes 2 and 3).

8. Kara White, "And Say the Cat Responded?," *Society & Animals* 21 (2013), 93–104.

9. Clinton R. Sanders and Arnold Arluke, "If Lions Could Speak: Investigating the Animal-Human Relationship and the Perspectives of Nonhuman Others," *Sociological*

Quarterly 34, no. 3 (1993): 377–390; Clinton Sanders, *Understanding Dogs* (Philadelphia: Temple University Press, 1999).

10. Claudio Ottoni, Wim Van Neer, Bea De Cupere, Julien Daligault, Silvia Guimaraes, Joris Peters, Nikolai Spassov, Mary E. Predergast, Nicole Boivin, and Arturo Morales-Muniz, "The Paleogenetics of Cat Dispersal in the Ancient World," *Nature Ecology & Evolution* 1 (2017): article 0139, online; Dennis C. Turner and Patrick Bateson, eds., *The Domestic Cat*, 3rd ed. (Cambridge: Cambridge University Press, 2014), 85–86.

11. Eva Jablonka and Marion J. Lamb, *Evolution in Four Dimensions: Genetic, Epigenetic, Behavioral, and Symbolic Variation in the History of Life* (Cambridge: MIT Press, 2014); Andràs Pàldi, *L'Épigenetique ou la nouvelle ère de l'hérédité* (Paris: Le Pommier, 2018); Ludovic Orlando, "L'AND ancient comme nouvelle source historique," in *Aux sources de l'histoire animale*, ed. Éric Baratay (Paris: Éditions de la Sorbonne, 2019), 199–210.

Chapter 2. House Cats and Garden Cats

1. François-Augustin de Moncrif, *Moncrif's Cats*, trans. Reginald Bretnor (New York: A. S. Barnes, 1965); Édouard Manet in *Les Chats*, by Champfleury (Paris: Rothschild, 1869); Théodore de Banville, "Le chat," in *Les Animaux chez eux*, by August Lançon (Paris: Baschet, 1882), 40; Stéphane Mallarmé, "Pauvre Enfant pâle," *Divagations* (Paris: Fasquelle, 1897), 15; Silvia Mergenthal, "Fremde im eigenen Haus: Die Katze in angloamerkanischen Schauerroman," and Ulrike Landfester, "Von Klassikern, Klugscheissern und Koautoren: Di Katze im Kriminalroman," in *Von Katzen und Menschen*, ed. Clemens Wischermann (Kostanz: UVK, 2007), 109–121, 123–137.

2. Georges Docquois, *Bêtes et gens de lettres* (Paris: Flammarion, 1896), 32 (quote 1), 77–78 (quote 2), 260–263 (quotes 3–4); Dennis C. Turner and Patrick Bateson, eds., *The Domestic Cat*, 3rd ed. (Cambridge: Cambridge University Press, 2014), 45.

The Jaunets

1. Jean-Henri Fabre, "Histoire de mes chats," chap. 8 in *Souvenirs entomologiques*, série 2 (1882; Paris: Delgrave, 1890–1900), 124–133; Madelaine Pinault-Soerensen, ed., *De l'Homme et des insectes: Jean-Henri Fabre, 1823–1915* (Paris: Somogy, 2003); Émile de la Bédollierre, *Histoire de la mère Michel et de son chat* (1846; Paris: Blanchard, 1853).

2. On an applied methodology see Éric Baratay, ed., *Aux Sources de l'histoire animale* (Paris: Editions de la Sorbonne, 2019); Baratay, ed., *Croiser les sciences pour lire les animaux* (Paris: Éditions de la Sorbonne, 2020); Baratay, ed., *Concepts animaux, déas-anthropisés, décloisonnés* (forthcoming).

3. Jean-Henri Fabre, *Souvenirs entomologiques* (Paris: Delgrave, 1890–1900), 125; Jean-M.-M. Rédarès, *Traité raisonné sur l'éducation du chat domestique* (Paris: Bourayne, 1835), 9, 12; Dennis C. Turner and Patrick Bateson, eds., *The Domestic Cat*, 3rd ed. (Cambridge: Cambridge University Press, 2014), 40, 48–51, 76.

4. Pierre Loti, "A Dying Cat," in *The Book of Pity and of Death*, trans. T. P. O'Connor (1891; New York: Cassell, 1892); Rédarès, *Traité raisonné sur l'education*, 100–108; Alexandre Landrin, *Le Chat* (Paris: Carré, 1894), 216–229.

5. Fabre, *Souvenirs*, 125–126; Susan Soennichsen and Arnold S. Chamove, "Response of Cats to Petting by Humans," *Anthrozoös* 15, no. 3 (2002): 258–265; Turner and Bateson, *Domestic Cat*, 39–45.

6. Turner and Bateson, *Domestic Cat*, 39–55; Janet M. and Steven F. Alger, *Cat Culture: The Social World of a Cat Shelter* (Philadelphia: Temple University Press, 2003).

7. Baratay, *Croiser les sciences.*

8. Fabre, *Souvenirs*, 125–136, 129; Landrin, *Le Chat*, 216–241; Louis-Georges Neumann, *Parasites et maladies parasitaires du chien et du chat* (Paris: Asselin, 1914); Joaquim P. Ferreira, Inês Leitão, Margarita Santos-Reis, and Eloy Revilla, "Human-Related Factors Regulate the Spatial Ecology of Domestic Cats in Sensitive Areas for Conservation," *PLoS One* 6, no. 10 (2011): e25970.

9. Turner and Bateson, *Domestic Cat*, 48–50; Francois-Augustin de Moncrif, *Moncrif's Cats*, trans. Reginald Bretnor (New York: A. S. Barnes, 1965), 101; Rédarès, *Traité raisonné sur l'education*, 26; Alain Rey, ed., *Dictionnaire historique de la langue française* (Paris: Robert, 1992), 2:1831; Théophile Gautier, *Ménagerie intime* (Paris: Lemerre, 1869), 20, 23; Pierre Loti, *Journal* (Paris: Les Indes Savants, 2012), 3:231.

10. Fabre, *Souvenirs*, 125–130; Frances Simpson, *The Book of the Cat* (London: Cassel, 1903), 18; Turner and Bateson, *Domestic Cat*, 41–54, 237; Éric Pierre, "La Zoophilie dans ses rapports à la philanthropie," *Cahiers d'histoire* 42, nos. 3–4 (1997): 655–675.

11. Fabre, *Souvenirs*, 129–130; E. Sanfourche, *Les Chiens, les chats, et les oiseaux* (Paris: Auteur, 1866), 90; Turner and Bateson, *Domestic Cat*, 134, 188–189, 207; Marcus Schneck Marcus Schneck and Jill Caravan, *Le Chat*, (Paris: Solar, 1991), 28; Kazuki Miyaji, Maki Kato, Nobuyo Ohtani, and Mitsuaki Ohta, "Experimental Verification of the Effects on Normal Domestic Cats by Feeding Prescription Diet for Decreasing Stress," *Journal of Applied Animal Welfare Science* 18, no. 4 (2015): 355–362; Marta Amat, Tomàs Camps, and Xavier Manteca, "Stress in Owned Cats: Behavioural Changes and Welfare Implications," *Journal of Feline Medicine and Surgery* 18, no. 8 (2016): 577–586.

12. Astrid Guillaume, "La sémantique et la sémiotique au service du langage animal," in *Regards sur l'animal et son langage*, ed. Sandra Contamina and Fernando Copello (Rennes: PUR, 2022).

13. Fabre, *Souvenirs*, 127–128; Rédarès, *Traité raisonné sur l'éducation*, 82–84; Turner and Bateson, *Domestic Cat*, 40, 45; Schneck and Caravan, *Le Chat*, 28; Éric Baratay, "Chacun jette son chien!" *Romantisme* 153 (2011): 147–162; Damien Baldin, "Éliminer le mauvais chien," in *Une Bête parmi les hommes: Le Chien*, ed. Fabrice Guizard and Corinne Beck (Amiens: Encrage, 2014), 403–415.

14. Fabre, *Souvenirs*, 129–131; Turner and Bateson, *Domestic Cat*, 134, 189, 205–207; Amat, Camps, and Manteca, "Stress in Owned Cats."

15. Turner and Bateson, *Domestic Cat*, 12, 21–22, 26, 116, 120; Randolph M. Baral, Navneet K. Dhand, Kathleen P. Freeman, Mark B. Krockenberger, and Merran Govendir, "Biological Variation and Reference Change Values of Feline Plasma Biochemistry Analytes," *Journal of Feline Medicine and Surgery* 16, no. 4 (2015): 317–325; Leslie A. Lyons, "DNA Mutations of the Cat," *Journal of Feline Medicine and Surgery* 17, no. 3 (2015): 203–219; Tamara Leonova, "L'Approche écologique de la cognition sociale et son impact sur la conception des traits de personnalité," *L'Année Psychologique* 104, no. 2 (2004): 249–294; Carla A. Litchfield, Gillian Quinton, Hayley Tindle, Belinda Chiera, K. Heidy Kikillus, and Philip Roetman, "The 'Feline Five': An Exploration of Personality in Pet Cats (*Felis catus*)," *PLoS ONE* 12, no. 8 (2017): e018345; Janice M. Siegford, "Validation of a Temperament Test for Domestic Cats," *Anthrozöos* 16, no. 4 (2003): 332–351; Dominique Lestel, *L'Animal singulier* (Paris: Seuil, 2004).

16. Éric Baratay, *Animal Biographies: Toward a History of Individuals*, trans. Lindsay Turner (Athens: University of Georgia Press, 2022); Baratay, "Penser les individus," in *Les études animales sont-elles bonnes à penser?* ed. Aurélie Choné, Isabelle Iribarren, Marie Pelé, Catherine Repussard, and Cédric Sueur (Paris: L'Harmattan, 2020), 55–79; Baratay, *Aux Sources de l'histoire animale.*

17. Fabre, *Souvenirs*, 130–132; D. H. Lawrence, *Lady Chatterley's Lover* (Mineola, N.Y.: Dover Thrift Editions, 2006), 47; Turner and Bateson, *Domestic Cat*, 239; Marie-Amélie Forin-Wiart, "Identification des facteurs de variation et de la prédation exercée par les chat domestiques (Felis silvstris catus) en milieu rural" (thesis in biology, Université de Reims Champagne-Ardenne, 2014), 23–24, 98–104.

Toto

1. Athénaïs Michelet, *Mes Chats* (1906; Rennes: La Part Commune, 2004), 174–177.

2. Jocelyne Porcher, "Une Sociologie des animaux au travail," in *Les Animaux: Deux ou trois choses que nous savons d'eux*, ed. Vinciane Despret and Raphaël Lerrère (Paris: Hermann, 2014), 100–114.

3. Athénaïs Michelet, *Mes Chats*, 178–185, Jean-M.-M. Rédarès, *Traité raisonné sur l'éducation du chat domestique* (Paris: Bourayne, 1835), 69–70, 73; Dennis C. Turner and Patrick Bateson, eds., *The Domestic Cat* (Cambridge: Cambridge University Press, 2014), 12–16, 23–24, 57, 78, 198; Mathilde Gaudry, "Contribution au traitement de l'obsésité chez le chat" (veterinarian thesis, Maisons-Alfort, 2014), 93; Marcus Schneck and Jill Caravan, *Le Chat* (Paris: Solar, 1991), 25; Marie-Amélie Forin-Wiart, "Identification des facteurs de variation et de la prédation exercée par les chat domestiques (Felis silvstris catus) en milieu rural" (thesis in biology, Université de Reims Champagne-Ardenne, 2014) 75; Kazuki Miyaji, Ai Kobayashi, Teppei Maruko, Nobuyo Ohtani, and Mitsuaki Ohta, "Acoustic Signals of a Dog and Cat Induce Hemodynamic Responses within the Human Brain," *Anthrozoös* 27, no. 2 (2014): 165–172.

Moumoutte Blanche

1. Jean-M.-M. Rédarès, *Traité raisonné sur l'éducation du chat domestique* (Paris: Bourayne, 1835), 40–41, 50–51; E. Sanfourche, *Les Chiens, les chats, et les oiseaux* (Paris: Auteur, 1866), 87–88; Théodore de Banville, "Le Chat," in *Les Animaux chez eux*, by August Lançon (Paris: Baschet, 1882), 40; Gaston Percheron, *Le Chat* (Paris: Didot, 1885), 79, 88, 93, 128–130; Alexandre Landrin, *Le Chat* (Paris: Carré, 1894), 27, 107, 113, 117, 124, 141–144; Miss Hugh Miller, *Cats and Dogs* (London: Nelson, 1868); Frances Simpson, *The Book of the Cat* (London: Cassel, 1903), 19, 22, 26, 30–31.

2. Georges Docquois, *Bêtes et gens de lettres* (Paris: Flammarion, 1896), 58–59; Pierre Loti, *The Book of Pity and of Death*, trans. T. P. O'Connor (1891; New York: Cassell, 1892), 5:116; Loti, *Journal: 1879–1886* (Paris: Les Indes Savantes, 2008), 2:9–23; Martine Sagaert, "Loti, la mer et la mère," and Rolande Leguillon, "Loti et la peur du néant," in *Loti en son temps* (Rennes: PUR, 1994), 143–151, 251–258; Alain Buisine, *Pierre Loti: L'Écrivain et son double* (Paris: Tallandier, 2004).

3. Loti, *Book of Pity*, 563, 579 (quote 1); Loti, *The Story of a Child*, trans. Carolyn F. Smith (Boston: C. C. Birchard, 1901), 6 (quote 2); Loti, *Prime Jeunesse* (Paris: Calmann-Lévy, 1919), 11–12, 23, 107–112, 156, 257; Loti, *Journal*, 2:195, 3:230–231; Percheron, *Le Chat*, 130; blueprint, old photographs and more recent photos can be found at the following website: maisondepierreloti.fr.

4. *Le Soir*, January 22, 1949, p. 16; Dennis C. Turner and Patrick Bateson, ed., *The Domestic Cat* (Cambridge: Cambridge University Press, 2014); Marcus Schneck and Jill Caravan, *Le Chat* (Paris: Solar, 1991), 40.

5. Jacques Pimpaneau, *Chine: Histoire de la littérature* (Arles: Picquier, 2016), 25, 30.

6. Loti, *Book of Pity*, 59, 93; Loti, *Story of a Child*, 4, 114, 182; Loti, *Prime Jeunesse*, 11–12; Loti, "Suleima," in *Fleurs d'ennui* (1882; Paris: Calmann-Lévy, 1926), 353 (quote 3), 354; Loti, *Journal*, 2:153, 529, 544, 553, 712, 714, 723, 750, 3:59; Turner and Bateson, *Domestic Cat*, 39–40.

7. Loti, *Book of Pity*; *Story of a Child*; *Prime Jeunesse*, 12–13, 225; Loti, *Journal*, 2:152, 543, 56, 707, 744.

8. In the English translation by T. P. O'Connor, this phrase is translated as "Be-off!" [T. N.].

9. Loti, *Book of Pity*, 103–105; *Story of a Child*; *Prime Jeunesse*; *Journal*, vol. 2.

10. Loti, *Book of Pity*, 57–58, 95.

11. Loti, *Story of a Child*, 9, 35–49; *Prime Jeunesse*, 35, 61.

12. Loti, *Book of Pity*, 95; *Prime Jeunesse*.

13. Loti, *Book of Pity*, 95, 58–59; *Prime Jeunesse*, 22; Émile de la Bedollière, *Le Chat de la mère Michel et de son chat* (1846; Paris: Blanchard, 1853), 19, 26; Rédarès, *Traité raisonné sur l'éducation*, 30; Dennis C. Turner, Gerulf Rieger, and Lorenz Gygaz, "Spouses and Cats and Their Effects on Human Mood," *Anthrozoös* 16, no. 3 (2003): 213–228; Ai Kobayashi, Yusuke Yamaguchi, Nobuyo Ohtani, and Mitsuaki Ohta, "The Effects of Touching and Stroking a Cat on the Inferior Frontal Gyrus in People," *Anthrozoös* 30, no. 3 (2017): 473–486.

14. Clinton R. Sanders, *Understanding Dogs* (Philadelphia: Temple University Press, 1999); Janet M. and Steven F. Alger, *Cat Culture: The Social World of a Cat Shelter* (Philadelphia: Temple University Press, 2003); David Goode, *Playing with My Dog Katie* (West Lafayette, Ind.: Perdue University Press, 2006); Samantha Hurn, *Humans and Other Animals* (London: Pluto Press, 2012); Marion Vicart, *Des Chiens auprès des hommes* (Paris: Petra, 2014); Marcus Baynes-Rock, *Among the Bone-Eaters* (University Park: Penn State University Press, 2015); John Bradshaw, *The Animal among Us* (London: Penguin, 2017).

15. Percheron, *Le Chat*, 129; Landrin, *Le Chat*; Rédarès, *Traité raisonné sur l'éducation*, 9, 12.

16. Loti, *Book of Pity*; *Story of a Child*; *Journal*, 2:488, 3:55; Loti, "Chiens et Chats," in *Reflets sur la sombre route (Eubres complètes)* (Paris: Calmann Lévy, 1893–1911), 8:471–473; Rédarès, *Traité raisonné sur l'éducation*, 80–81; Turner and Bateson, *Domestic Cat*, 73, 75, 117; Beatrice Alba and Nick Haslam, "Dog People and Cat People Differ on Dominance-Related Traits," *Anthrozoös* 28, no. 1 (2015): 37–44.

17. Loti, *Journal*, 2:489, 522–523, 529, 539, 543–544, 557, 604–605, 658, 701, 706–707, 712, 730, 744, 3:140, 152, 163.

18. Loti, *Book of Pity*, 101; *Journal*, 2:702, 705–707, 720; *Story of a Child*.

19. Loti, *Book of Pity*; *Journal*, 2:702, 705–707, 720; "Suleïma," 308, 351, 353; *Story of a Child*; Rédarès, *Traité raisonné sur l'éducation*, 84; Percheron, *Le Chat*, 185–186; Turner and Bateson, *Domestic Cat*, 12–19, 28–31, 74, 79; Oxána Bánszegi, Peter Szenczi, Andrea Urritia, and Robyn Hudson, "Conflict or Consensus? Synchronous Change in Mother-Young Vocal Communication across Weaning in the Cat," *Animal Behavior* 130 (2017): 233–240.

20. Loti, *Book of Pity*, 90, 95, 105, 118; *Journal*, 2:701, 777, 3:147, 160, 177, 187, 234; Turner and Bateson, *Domestic Cat*, 41–46, 251.

21. Loti, *Book of Pity*; *Journal*, 3:59–60; Turner and Bateson, *Domestic Cat*, 19, 75, 115; Margaret Slater, Laurie Garrison, Katherine Miller, Emily Weiss, Kathleen Makolinski, Natasha Drain, and Alex Mirontshuk, "Practical Physical and Behavioral Measures to Assess the Socialization Spectrum of Cats," *Animals* 3, no. 4 (2013): 1162–1193.

22. Loti, *Book of Pity*, 108–109; *Journal*, 3:228, 230–231, 234; Rédarès, *Traité raisonné sur l'éducation*, 98–107; Landrin, *Le Chat*, 208–211; E. Despret, "Les Maladies infectieuses du chats" (veterinarian thesis, Maisons-Alfort, 1937); Docquois, *Bêtes et gens de lettres*, 69; Éric Baratay, "Chacun jette son chien!" *Romantisme* 153 (2011): 147–162.

Chapter 3. Inside Cats

1. Athénaïs Michelet, *Mes chats* (1906; Rennes: La Part Commune, 2004), 171–173, 218.

2. Mary Dickens, *My Father as I Recall Him* (London: Roxburghe Press, 1896).

3. Georges Docquois, *Bêtes et gens de lettres* (Paris: Flammarion, 1896), 35–43; Robert Mitchell, "Americans' Talk to Dogs: Similarities and Differences with Talk to Infants," *Research on Language & Social Interaction* 34, no. 2 (2001): 183–210; Dennis Burnham, Elizabeth Francis, U. Vollmer-Conna, C. Kitamura, Vicky Averkiou, A. Olley, M. Nguyen, and Cal Paterson, "Are You My Little Pussy-Cat? Acoustic, Phonetic, and Affective Qualities of Infant- and Pet-Directed Speech," 5th International Conference on Spoken Language Processing, Sydney, 1998, 4534–4556.

4. Athénaïs Michelet, *Mes chats*, 29–59.

Pierrot

1. Paule Petitier, "Chimères de compagnie," in *Ménagerie intime*, by Théophile Gautier (Paris: Éditions des Equateurs, 2008); Christine Majeune-Girodias, "La 'Menagerie intime' de Théophile Gautier," in *Histoire(s) et enchantement*, ed. Pascale Auraix-Jonchière (Clermont-Ferrand: PU Blaise-Pascal, 2009), 181–195; Stéphane Guégan, *Théophile Gautier* (Paris: Gallimard, 2011). For a very detailed chronology see the website including correspondence at www.theophilegautier.fr.

2. Judith Gautier, *Le Collier des jours* (Paris: Juven, 1904), and *Le Second Rang du collier* (Paris: Juven, s.d. 1905–1908).

3. Théophile Gautier, *Ménagerie intime* (Paris: Lemerre, 1869), 17–18; Judith Gautier, *Le Collier des jours*, 264; Georges Docquois, *Bêtes et gens de lettres* (Paris: Flammarion, 1896), 54.

4. Claudia Mertens, "Human-Cat Interactions in the Home Setting," *Anthrozöos* 4, no. 4 (1991): 214–231; Manuel Wedl, Barbara Bauer, Dorothy Gracey, Christine Grabmeyer, Elisabeth Spielauer, Jon Day, and Kurt Kotrschal, "Factor Influencing the Temporal Patterns of Dyadic Behaviors and Interactions between Domestic Cats and Their Owners," *Behavioral Processes* 86, no. 1 (2011): 58–67; Dennis C. Turner and Patrick Bateson, eds., *The Domestic Cat* (Cambridge: Cambridge University Press, 2014), 115–126

5. Judith Gautier, *Le Collier des jours*, 122, 235, 238, 244, 250; Judith Gautier, *Le Second Rang du collier*, 5, 7, 11; Théophile Gautier, *Ménagerie intime*, 18.

6. Jean-Michel Roubineau, *Les Cités grecques: Essai d'histoire sociale* (Paris: PUF, 2015)

7. Jean-Claude Filloux, *Psychologie des animaux* (Paris: PUF, "Que sais-je?," 1950).

8. Émile Benveniste, "Animal Communication and Human Language," in *Problems in General Linguistics*, trans. Mary Elisabeth Meeks (Carol Gables, Fla.: University of Miami Press, 1971), 49; Guillaime Lecointre and Hervé Le Guyader, *Classification phylogénétique du vivant* (Paris: Belin, 2006); Éric Baratay, "Les Dessous de la personalité non-humaine," in *La Personnalité juridique de l'animal*, ed. Caroline Regad and Cédric Riot (Paris: LexisNexis, 2020), 2:11–25; Éric Baratay, ed., *Concepts animaux, désanthropisés, décloisonnés* (forthcoming); Astrid Guillaume, "La Sémantique et la sémiotique au service du langage animale," in *Regards sur l'animal et son langage*, ed. Sandra Contamina and Fernando Copello (Rennes: PUR, 2022); Janet M. and Steven F. Alger, "Beyond Mead: Symbolic Interaction Between Humans and Felines," *Society and Animals* 5, no. 1 (1997): 65–81; Clinton R. Sanders, "Actions Speak Louder Than Words," *Symbolic Interaction* 26, no. 3 (2003): 405–426.

9. Théophile Gautier, *Ménagerie intime*, 18, 23; Jean-M.-M. Rédarès, *Traité raisonné sur l'éducation du chat domestique* (Paris: Bourayne, 1835), 72; Judith Gautier, *Le Collier des jours*, 109; Turner and Bateson, *Domestic Cat*, 118; Cara M. Hansen, "Movements and Predation Activity of Feral and Domestic Cats (Felis catus) on Banks Peninsula (PhD dissertation, Christchurch Lincoln University, 2010).

10. Judith Gautier, *Le Collier des jours*, 248, 254, 279, 281; Judith Gautier, *Le Second Rang du collier*, 2, 4–5, 8, 16; Marcus Schneck and Jill Caravan, *Le Chat* (Paris: Solar, 1991), 18; *Élever et soigner son chat* (Paris: Larousse, 2015), 38–39.

11. Théophile Gautier, *Ménagerie intime*, 18, 19, 22–23l; Pierre Loti, *The Book of Pity and of Death*, trans. T. P. O'Connor (1891; New York: Cassell, 1892); Turner and Bateson, *Domestic Cat*, 39–41, 45, 51, 194–199, 238; Claude Béata, ed., *L'Attachement* (Marseille: Solal, 2009); Marlitt Wendt, *Mieux comprendre son cheval* (Paris: Vigot, 2013), 49, 54; Marion Vicart, *Des Chiens auprès des hommes* (Paris: Editions Petra, 2014).

12. Théophile Gautier, *Ménagerie intime*, 19–20, 21–22; Turner and Bateson, *Domestic Cat*, 39–46, 75–78; Zana Bahlig-Pieren and Dennis C. Turner, "Anthropomorphic Interpretations and Ethological Descriptions of Dog and Cat Behavior by Lay People," *Anthrozoös* 12, no. 4 (1999): 205–210; Sarah L. H. Ellis, Victoria Swindell, and Oliver H. P. Burman, "Human Classification of Context-Related Vocalizations Emitted by Familiar and Unfamiliar Domestic Cats," *Anthrozoös* 28, no. 4 (2015): 625–634.

13. Jocelyne Porcher, "Une Sociologie des animaux au travail," in *Les Animaux: Deux ou trois choses que nous savons d'eux*, ed. Vinciane Despret and Raphaël Lerrère (Paris: Hermann, 2014); Turner and Bateson, *Domestic Cat*, 115, 117.

14. Judith Gautier, *Le Collier des jours*, 264; Rédarès, *Traité raisonné sur l'éducation*, 36, 39; Théophile Gautier, *Ménagerie intime*, 22; Docquois, *Bêtes et gens de lettres*, 94, 184–188.

15. Melissa R. Shyan-Norwalt, "Caregiver Perceptions of What Indoor Cats Do 'for Fun,'" *Journal of Applied Animal Welfare Science* 8, no. 3 (2005): 199–209; Colleen Wilson, Melissa Bain, and Gary Landsberg, "Owner Observations Regarding Cat Scratching Behavior," *Journal of Feline Medicine and Surgery* 18, no. 10 (2016): 791–797; Turner and Bateson, *Domestic Cat*, 54, 78, 188, 238; Théophile Gautier, *Ménagerie intime*, 18, 23; Schneck and Caravan, *Le Chat*, 25; Judith Gautier, *Le collier des jours*, 235, 279, 281; Judith Gautier, *Le Second Rang du collier*, 6, 15, 16; Rédarès, *Traité raisonné sur l'éducation*, 14, 76–79; Pete Biro and J. Stamps, "Are Animal Personality Traits Linked to Life-History

Productivity?" *Trends in Ecology & Evolution* 23, no. 7 (2008): 361–368; Bahlig-Pieren and Turner, "Anthropomorphic Interpretations and Ethological Descriptions"; Janet M. and Steven F. Alger, *Cat Culture: The Social World of a Cat Shelter* (Philadelphia: Temple University Press, 2003), 88, 90.

16. Théophile Gautier, *Ménagerie intime*, 29–31, 35–37; Judith Gautier, *Le Collier des jours*, 22, 24, 28, 32, 35, 37, 40, 43, 55, 80, 96, 115, 180, 227–228, 277; Turner and Bateson, *Domestic Cat*, 17.

17. Émile Zola, "Le Paradis des chats," in *Nouveau Centre à Ninon* (Paris: Charpentier, 1893), 78–76; Docquois, *Bêtes et gens de lettres*, 63–66.

Part 2. Possibles

1. Gaston Percheron, *Le Chat* (Paris: Didot, 1885) 149; Jean-M.-M. Rédarès, *Traité raisonné sur l'éducation du chat domestique* (Paris: Bourayne, 1835), 16, 18–19, 31, 36–37.

Chapter 4. Conversions

1. Jean-M.-M. Rédarès, *Traité raisonné sur l'éducation du chat domestique* (Paris: Bourayne, 1835), 7–80, 88–89; Christopher Smart, *Jubilate Agno*, published under the title *Rejoice in the Lamb: A Song from Bedlam*, edited by William Stead (New York: Henry Holt, 1939).

Trim

1. Claudio Ottoni, Wim Van Neer, Bea De Cupere, Julien Daligault, Silvia Guimaraes, Joris Peters, Nikolai Spassov, Mary E. Predergast, Nicole Boivin, and Arturo Morales-Muniz, "The Paleogenetics of Cat Dispersal in the Ancient World," *Nature Ecology & Evolution* 1 (2017): article 0139, online.

2. Harold M. Cooper, "Matthew Flinders (1774–1814)," in *Australian Dictionary of Biography* (Melbourne: Melbourne University Press, 1966), 1:389–391; numerous biographies such as the one by Ernest Scott, *The Life of Matthew Flinders* (Sydney: Angus, & Robertson, 1914) and more recently by Kenneth Morgan, *Matthew Flinders* (Sydney: Bloomsbury, 2017).

3. Matthew Flinders, *A Biographical Tribute to the Memory of Trim*, viewable online at https://en.wikisource.org/wiki/A_biographical_tribute_to_the_memory_of_Trim, paragraphs 1, 22, 26. Publications: Matthew Flinders, *Trim* (Sydney: Collins, 1977; Pymble, Australia: Angus and Robertson, 1997); Éric Baratay, *Animal Biographies: Toward a History of Individuals*, trans. Lindsay Turner (Athens: University of Georgia Press, 2022).

4. "Biographical Memoir of Captain Mathew Flinders, R.N.," *Naval Chronicle* 32 (1814): 177–191; Matthew Flinders, *A Voyage to Terra Australis* (London: Nicol, 1814).

5. Flinders, *Voyage to Terra Australis*, 1:civ; Scott, *Life of Captain Matthew Flinders*, 95–96; Flinders, *Biographical Tribute to the Memory*; Dennis C. Turner and Patrick Bateson, eds., *The Domestic Cat* (Cambridge: Cambridge University Press, 2014), 12–18, 57; Oxàna Bánszegi, Peter Szenczi, Andrea Urritia, and Robyn Hudson, "Conflict or Consensus? Synchronous Change in Mother-Young Vocal Communication across Weaning in the Cat," *Animal Behavior* 130 (2017): 233–240.

6. Flinders, *Biographical Tribute to the Memory*; Turner and Bateson, *Domestic Cat*, 13, 16, 18, 239; Margaret Slater, Laurie Garrison, Katherine Miller, Emily Weiss, Kathleen Makolinski, Natasha Drain, and Alex Mirontshuk, "Practical Physical and Behavioral Measures to Assess the Socialization Spectrum of Cats," *Animals* 3, no. 4 (2013): 1162–1193.

7. Flinders, *Biographical Tribute to the Memory*; Turner and Bateson, *Domestic Cat*, 39, 52–53, 258; Florence Burgat, *Vivre avec un inconnu* (Paris: Rivages Poche, 2016), 38, 43–44; Jean-François Bert, "Lire ce que Marcel Mauss a lu," *Le Portique* 17 (2006), online.

8. Marcel Mauss, "Les Techniques du corps," *Journal de la Psychologie* 32, nos. 3–4 (1936): 271–293; Eduardo Kohn, *How Forests Think: Toward an Anthropology beyond the Human* (Berkeley: University of California Press, 2013).

9. Flinders, *Biographical Tribute to the Memory*; Turner and Bateson, *Domestic Cat*, 21–22, 48–50, 80, 116, 120, 238; Carla A. Litchfield, Gillian Quinton, Hayley Tindle, Belinda Chiera, K. Heidy Kikillus, and Philip Roetman, "The 'Feline Five': An Exploration of Personality in Pet Cats (*Felis catus*)," *PLoS ONE* 12, no. 8 (2017): e018345; Sarah Lowe and John W. S. Bradshaw, "Responses of Pet Cats to Being Held by an Unfamiliar Person," *Anthrozoös*, 15, no. 1 (2002): 69–79; Dominique Lestel, *L'Animal singulier* (Paris: Seuil, 2004); Lestel, *Les Amis de mes amis* (Paris: Seuil, 2007); Donna Haraway, *The Companion Species Manifesto: Dogs, People, and Significant Otherness* (Chicago: Prickly Paradigm Press, 2003).

10. Flinders, *Biographical Tribute to the Memory*; Turner and Bateson, *Domestic Cat*, 68, 198; Marie-Amélie Forin-Wiart, "Identification des facteurs de variation et de la prédation exercée par les chat domestiques (Felis silvstris catus) en milieu rural" (thesis in biology, Université de Reims Champagne-Ardenne, 2014), 103–104, 125; Pete Biro and J. Stamps, "Are Animal Personality Traits Linked to Life-History Productivity?" *Trends in Ecology & Evolution* 23, no. 7 (2008): 361–368; Laurence Sterne, *The Life and Opinions of Tristram Shandy, Gentleman* (London: Ward, Dodsley, Beckett & DeHondt, 1759–1767).

11. Erving Goffman, *Interaction Ritual: Essays in Face-to-Face Behavior* (New York: Pantheon, 1982); Janet M. and Steven F. Alger, "Cat Culture, Human Culture," *Society & Animals* 7, no. 3 (1999): 199–218; Janet M. and Steven F. Alger, *Cat Culture: The Social World of a Cat Shelter* (Philadelphia: Temple University Press, 2003); Florent Kohler, "Blondes d'Aquitaine," *Études rurales* 189 (2012): 155–174; Jocelyne Porcher, "Une Sociologie des animaux au travail," in *Les Animaux: Deux ou trois choses que nous savons d'eux*, ed. Vinciane Despret and Raphaël Lerrère (Paris: Hermann, 2014).

12. Flinders, *Biographical Tribute to the Memory*.

13. Keith Thomas, *Man and the Natural World: Changing Attitudes in England 1500–1800* (New York: Penguin, 1985).

14. Flinders, *Biographical Tribute to the Memory*; Marcus Schneck and Jill Caravan, *Le Chat* (Paris: Solar, 1991), 18, 28, 36; Jean-M.M. Rédarès, *Traité raisonné sur l'éducation du chat domestique* (Paris: Bourayne, 1835), 68–70; Christopher Smart, *Jubilate Agno*, "Fragment B [For I Will Consider my Cat Jeoffry]," 1759–1763, online; David Goode, *Playing with My Dog Katie* (West Lafayette, Ind.: Perdue University Press, 2006); Donna Haraway, "Jeux de ficelles avec les espèces compagnes," in Despret and Lerrère, *Les Animaux*, 23–59.

15. Flinders, *Biographical Tribute to the Memory*; Flinders, *Voyage to Terra Australis*, 1:cxx, cxxviii, cxciv; Scott, *Life of Captain Matthew Flinders*, 95–96, 123–140, 158–163.

16. Flinders, *Biographical Tribute to the Memory*; Flinders, *Voyage to Terra Australis*, 1:4–5,15–18, 21, 28; Scott, *Life of Captain Matthew Flinders*, 174–176.

17. Flinders, *Biographical Tribute to the Memory*; Dany Bréelle, "Matthew Flinders et la mise en cartes d'un nouvel espace indo-pacifique," *Cybergeo*, document 797 (2016), online.

Chapter 5. Amplification

1. Donna Haraway, "Jeux de ficelle avec les espèces compagnes," in *Les Animaux: deux ou trois choses que nous savons d'eux*, ed. Vinciane Despret and Raphaël Lerrère (Paris: Hermann, 2014), 23–59; Dennis C. Turner and Patrick Bateson, eds., *The Domestic Cat* (Cambridge: Cambridge University Press, 2014), 19–20, 26.

Minette

1. Athénaïs Michelet, *Mes chats* (1906; Rennes: La Part Commune, 2004), 30–33, 48; Eugène Noël, *Michelet et ses enfants* (Paris: Dreyfus, 1878), 207–208; Marcus Schneck and Jill Caravan, *Le Chat* (Paris: Solar, 1991), 25; Dennis C. Turner and Patrick Bateson, eds., *The Domestic Cat* (Cambridge: Cambridge University Press, 2014), 114.

2. Athénaïs Michelet, *Mes Chats*, 29–31, 33–34, 36–40; Schneck and Caravan, *Le Chat*, 28; Marta Armat, Tomàs Camps, and Xavier Manteca, "Stress in Owned Cats: Behavioural Changes and Welfare Implications," *Journal of Feline Medicine and Surgery* 18, no. 8 (2016): 577–586.

3. Athénaïs Michelet, *Mes Chats*, 35, 38, 40; Noël, *Michelet et ses enfants*, 207; Mme Jules Michelet, *Mémoires d'une enfant* (Paris: Hachette, 1867), 345, 388; Turner and Bateson, *Domestic Cat*, 18; James J. Gibson, *The Ecological Approach to Visual Perception* (New York: Psychology Press, 2015); Francisco Varela, Evan T. Thompson, and Eleanor Rosch, *Embodied Mind: Cognitive Science and Human Experience* (Cambridge: MIT Press, 1991).

4. Erving Goffman, *Interaction Ritual: Essays in Face-to-Face Behavior* (New York: Pantheon Books, 1982); Goffman, *The Presentation of Self in Everyday Life* (New York: Anchor Books, 1959)

5. Athénaïs Michelet, *Mes Chats*, 41–43, 47–49; Marcel Mauss, "Les Techniques du corps," *Journal de la psychologie* 32, nos. 3–4 (1936): 271–293; Jean-M.-M. Rédarès, *Traité raisonné sur l'éducation du chat domestique* (Paris: Bourayne, 1835), 9, 12, 36, 84; Gaston Percheron, *Le Chat* (Paris: Didot, 1885), 185–186; Mme Jules Michelet, *Mémoires d'une enfant*, 180, 241, 297; Turner and Bateson, *Domestic Cat*, 76.

6. Athénaïs Michelet, *Mes Chats*, 42–43, 50–51, 54–55, 59; Rédarès, *Traité raisonné sur l'éducation*, 80–81, 88; Percheron, *Le Chat*, 183; Ai Kobayashi, Yusuke Yamaguchi, Nobuyo Ohtani, and Mitsuaki Ohta, "The Effects of Touching and Stroking a Cat on the Inferior Frontal Gyrus in People," *Anthrozoös* 30, no. 3 (2017): 473–486; Jérôme Michalon, *Panser avec les animaux* (Paris: Presses des Mines, 2014).

7. Athénaïs Michelet, *Mes Chats*, 52, 61–63, 66–68, 72–74.

8. Athénaïs Michelet, *Mes Chats*, 75, 85–86, 93, 94–96, 97–100, 214; Kobayashi et al., "Effects of Touching and Stroking"; Dennis C. Turner, Gerulf Rieger, and Lorenz Gygaz,

"Spouses and Cats and Their Effects on Human Mood," *Anthrozoös* 16, no. 3 (2003): 213–228.

9. Elfriede Hermann, "Communicating with Transculturation," *Journal de la Société des Océanistes* 125 (2007): 257–260. Concerning the similar concept of interculturation, see Zobra Guerraoui, "De l'Acculturation à l'interculturation: Réflexions épistemologiques," *L'Autre* 10, no. 2 (2009): 195–200.

10. Marie-Amélie Forin-Wiart, "Identification des facteurs de variation et de la prédation exercée par les chat domestiques (Felis silvstris catus) en milieu rural" (thesis in biology, Université de Reims Champagne-Ardenne, 2014), 75.

11. G. Hasse, *Nos Chats* (Anvers: Koch, s.d., [1920]), 48; E. Larieux and Ph. Jumaud, *Le Chat: Races—elevage—Maladies* (Paris: Masson, 1927), 3, 6–7; M. Boilève de la Gombaudière, *Nos Compagnons . . . les chats* (Paris: Nilsson, 1932), viii, 17.

12. Larieux and Jumaud, *Le Chat*, 33; Adrien Loir, *Le Chat, son utilisé* (Paris: Baillière, 1930); Fernand Méry, ed., *Le Chat* (Paris: Larousse, 1973), 184–189.

13. Victor Rabaudy, *Le Chat, ce compagnon pafrois dangereux* (Toulouse: Imprimerie ouvrière, 1944), 41–47; Méry, *Le Chat*, 328; Clemens Wishcermann, ed., *Von Katzen und Menschen* (Kostanz: UVK, 2007), 155–171, 237–247.

14. François Héran, "Les Animaux domestiques," *Données Sociales* (1987): 417–423; Héran, "Comme chiens et chats," *Ethnologie Française* 18, no. 2 (1988): 325–337.

15. Wishcermann, *Von Katzen und Menschen*, 155–171, 211–224; Méry, *Le Chat*, 325–333; Turner and Bateson, *Domestic Cat*, 72, 137–141, 216; John C. New Jr., M. D. Salman, Mike King, Janet M. Scarlett, Philip H. Kass, and Jennifer M. Hutchison, "Characteristics of Shelter-Relinquished Animals and Their Owners Compared with Animals and Their Owners in U.S. Pet-Owning Households," *Journal of Applied Animal Welfare Science* 3, no. 3 (2000): 179–201; Rachel A. Casey, Sylvia Vandenbussche, John W. S. Bradshaw, and Margaret A. Roberts, "Reasons for Relinquishment and Return of Domestic Cats (*Felis Sylvestris Catus*) to Rescue Shelters in the U.K.," *Anthrozoös* 22, no. 4 (2009): 347–358; Mathieu Harel, Mathilde Goujon, and Dominique Grandjean, "Contribution à l'étude des refuges félins en France" (veterinarian thesis, Maisons-Alfort, 2014); Sarah Zito, John Morton, Dianne Vankan, Mandy Paterson, Pauleen C Bennett, Jacquie Rand, and Clive J. C. Phillips, "Reasons People Surrender Unowned and Owned Cats to Australian Animal Shelters and Barriers to Assuming Ownership of Unowned Cats," *Journal of Applied Animal Welfare Science* 19, no. 3 (2016): 303–319.

Part 3. The Rise of Anthroposized Cultures

1. E. Larieux and Ph. Jumaud, *Le Chat: Races—elevage—maladies* (Paris: Masson, 1927), 133–136; Fernand Méry, ed., *Le Chat* (Paris: Larousse, 1973), 156–159; 287–295, 306–318, 337–340; *Élever et soigner son chat* (Paris: Larousse, 2015).

Chapter 6. Cats as Friends

Miton

1. Marie Dormoy, *Le Chat Miton*, preface by Paul Léautaud (Paris: Editions Spirales, s.d., 1948); *Chats par Ylla*, presented by Paul Léautaud (Paris: Éditions O.E.T., s.d., 1937); books on Ylla: *Ylla* (Chalon-Sur-Saône: Musée Nicéphore-Niepce, 1983); Paul

Léautaud, *Journal littéraire* (Paris: Mercure de France, 1986), 3:1665–1666, 1688, 1701–1702, which doesn't mention Ylla but the book is dedicated to Miton; Sévérine, *Sac-à-tout: Mémoires d'un petit chien* (Paris: Juven, 1903).

2. Dormoy, *Le Chat Miton*, 3, 7–11; Dennis C. Turner and Patrick Bateson, eds., *The Domestic Cat* (Cambridge: Cambridge University Press, 2014), 12–23, 57.

3. Dormoy, *Le Chat Miton*, 2, 12, 19, 23.

4. Jean-M.-M. Rédarès, *Traité raisonné sur l'éducation du chat domestique* (Paris: Bourayne, 1835), 89; Alexandre Landrin, *Le Chat* (Paris: Carré, 1894), 250; A. Rome, "De l'Ovariectomie de la chienne et du chat" (veterinarian thesis, Maisons-Alfort, 1927); R. Sauvaitre, "Contribution à l'étude de l'anesthésie générale par le somnifère chez le chien et le chat" (veterinarian thesis, Toulouse, 1929).

5. Dormoy, *Le Chat Miton*, 2–3, 22–26.

6. Ibid., 13–20, 22–24, 27; E. Larieux and Ph. Jumaud, *Le Chat: Races—elevage—maladies* (Paris: Masson, 1927), 133–136.

7. Turner and Bateson, *Domestic Cat*, 72–78; Melissa R. Shyan-Norwalt, "Caregiver Perceptions of What Indoor Cats Do 'for Fun,'" *Journal of Applied Animal Welfare Science* 8, no. 3 (2005): 199–209; Erving Goffman, *The Presentation of Self in Everyday Life* (New York: Anchor, 1959).

8. Dormoy, *Le Chat Miton*, 13–14, 23–27; Turner and Bateson, *Domestic Cat*, 40–53, 238.

9. Dormoy, *Le Chat Miton*, 2, 9–12, 14–15, 26–27.

10. Turner and Bateson, *Domestic Cat*, 142, 144, 160; Landrin, *Le Chat*, 249.

11. Dormoy, *Le Chat Miton*, 3–4, 7, 12, 26; Léautaud, *Journal littéraire*, 3:643; Fernand Méry, ed., *Le Chat* (Paris: Larousse, 1973), 306, 308; Marcus Schneck and Jill Caravan, *Le Chat* (Paris: Solar, 1991), 25; Marta Amat, Tomàs Camps, and Xavier Manteca, "Stress in Owned Cats: Behavioural Changes and Welfare Implications," *Journal of Feline Medicine and Surgery* 18, no. 8 (2016): 577–586; Turner and Bateson, *Domestic Cat*, 132–134, 152, 191, 195–196, 202; Andy Sparkes and Margie Scherk, "In Pursuit of Better Care, Compassion and Understanding of Cats Globally," *Journal of Feline Medicine and Surgery* 20, no. 1 (2018): 3.

Chapter 7. Cats as Family

1. Yves Navarre, *Une Vie de chat* (Paris: Albin Michel, 1986); Louis Nucéra, *Sa Majesté le chat* (Paris: L'Archipel, 1992).

2. Éric Baratay, *Animal Biographies: Toward a History of Individuals*, trans. Lindsay Turner (Athens: University of Georgia Press, 2022).

3. William S. Burroughs, *The Cat Inside* (New York: Grenfell, 1986).

Dewey

1. Vicki Myron and Brett Witter, *Dewey: The Small-Town Library Cat Who Touched the World* (New York: Grand Central Publishing, 2008).

2. Vicki Myron and Brett Witter, *Dewey's Nine Lives* (New York: Simon & Shuster, 2010); Myron and Witter, *Dewey: There's a Cat in the Library!* (New York: Little, Brown, 2009).

3. Myron and Witter, *Dewey*, 9–25; Marcus Schneck and Jill Caravan, *Le Chat* (Paris:

Solar, 1991), 26, 28; Dennis C. Turner and Patrick Bateson, eds., *The Domestic Cat* (Cambridge: Cambridge University Press, 2014), 48, 238.

4. Myron and Witter, *Dewey*, 80, 142–143, 160–161; Turner and Bateson, *Domestic Cat*, 12–23, 58.

5. James J. Gibson, *The Ecological Approach to Visual Perception* (New York: Psychology Press, 2014); Francisco Varela, Evan T. Thompson, and Eleanor Rosch, *Embodied Mind: Cognitive Science and Human Experience* (Cambridge: MIT Press, 1991); "La Place de l'animal," special issue of *Éspaces et Sociétés* 110–111 (2002).

6. Myron and Witter, *Dewey*, 89, 96–108.

7. Ibid., 6, 26–30, 31–32, 36, 50–56, 124; Turner and Bateson, *Domestic Cat*, 45, 54, 78, 144, 160, 197, 237; Schneck and Caravan, *Le Chat*, 25; Marcel Mauss, "Les Techniques du corps," *Journal de la Psychologie* 32, nos. 3–4 (1936): 271–293.

8. Myron and Witter, *Dewey*, 32–33, 37, 110.

9. Turner and Bateson, *Domestic Cat*, 23, 45, 54, 78, 116–117, 196–197, 237; Susan Soennichsen and Arnold S. Chamove, "Responses of Cats to Petting by Humans," *Anthrozoös* 15, no. 3 (2002): 258–265; Schneck and Caravan, *Le Chat*, 18, 36; *Élever et soigner son chat* (Paris: Larousse, 2015), 38–39; David Goode, *Playing with My Dog Katie* (West Lafayette, Ind.: Purdue University Press, 2006); Sarah E. Lowe and John W. S. Bradshaw, "Responses of Pet Cats to Being Held by an Unfamiliar Person," *Anthrozoös* 15, no. 1 (2002): 69–79.

10. Myron and Witter, *Dewey*, 28, 31, 109, 135, 164, 172; Marta Borgi and Francesca Cirulli, "Children's Preferences for Infantile Features in Dogs and Cats," *Human-Animal Interaction Bulletin* 1, no. 2 (2013): 1–15; Turner and Bateson, *Domestic Cat*, 76, 114, 197.

11. Myron and Witter, *Dewey*, 5, 12, 36–37; Turner and Bateson, *Domestic Cat*, 116, 160; Sarah L. H. Ellis, Victoria Swindell, and Oliver H. P. Burman, "Human Classification of Context-Related Vocalizations Emitted by Familiar and Unfamiliar Domestic Cats," *Anthrozoös* 28, no. 4, 2015, 625–634; Zana Bahlig-Pieren and Dennis C. Turner, "Anthropomorphic Interpretations and Ethological Descriptions of Dog and Cat Behavior by Lay People," *Anthrozoös* 12, no. 4 (1999): 205–210; Eduardo Kohn, *How Forests Think: Toward an Anthropology beyond the Human* (Berkeley: University of California Press, 2013).

12. Valérie Charasse, "Évolution de la place du chat au sein du foyer: Impact sur son niveau de médicalisation" (veterinarian thesis, Lyon, 2015), 117–119; Jacqueline M. Fantuzzi, Katherine A. Miller, and Emily Weiss, "Factors Relevant to Adoption of Cats in an Animal Shelter," *Journal of Applied Animal Welfare Science* 13, no. 2 (2010): 174–179; Mathieu Harel, Mathilde Goujon, and Dominique Grandjean, "Contribution à l'études des refuges félins en France" (veterinarian thesis, Maisons-Alfort, 2014), 135; Schneck and Caravan, *Le Chat*, 14–15, 18, 22–23, 36; *Élever et soigner son chat*, 38–39; Turner and Bateson, *Domestic Cat*, 196–197; Concerning dogs, see Éric Baratay, *Le Point de vue animal* (Paris: Seuil, 2012) and *Animal Biographies*, trans. Lindsey Turner (Athens: University of Georgia Press, 2022).

13. Myron and Witter, *Dewey*, 86–87, 112–113.

14. Ibid., 20–21, 50–54, 72–75; Turner and Bateson, *Domestic Cat*, 131, 188; Valarie V. Tynes, Leslie Sinn, and Colleen S. Koch, "The Relationship between Physiology and Behavior in Dogs and Cats," in Emily Weiss, Heather Mohan-Gibbons, and Stephen Zaw-

istowski, eds., *Animal Behavior for Shelter Veterinarians and Staff* (Hoboken, N.J.: Wiley, 2015), 63–100.

15. Myron and Witter, *Dewey*, 63; Turner and Bateson, *Domestic Cat*; Donald C. Turner, Gerulf Rieger, and Lorenz Gygaz, "Spouses and Cats and Their Effects on Human Mood," *Anthrozoös* 16, no. 3 (2003): 213–228; Jérôme Michalon, *Panser avec les animaux* (Paris: Presses des Mines, 2014): Ai Kobayashi, Yusuke Yamaguchi, Nobuyo Ohtani, and Mitsuaki Ohta, "The Effects of Touching and Stroking a Cat on the Inferior Frontal Gyrus in People," *Anthrozoös* 30, no. 3 (2017): 473–486.

16. Myron and Witter, *Dewey*, 22, 118–120, 154–155, 203–204; Éric Baratay and Philippe Delisle, eds., *Milou, Idéfix & Cie: Le Chien en BD* (Paris: Karthala, 2012).

17. Valérie Charasse, "Évolution de la place du chat au sein du foyer: Impact sur son niveau de médicalisation" (veterinarian thesis, Lyon, 2015), 117, 119; Ludovic Freyburger, ed., "La Médécine preventative du chien et du chat," special issue *PratiqueVet*, 2016; "Médécines régénératives chez le chien et le chat," *Le Point vétérninaire* canine expert, 371 (2016): 21–43.

18. Jan Bellows, Sharon Center, Leighann Daristotle, Amara H. Estrada, Elizabeth A. Flickinger, Debra F. Horwitz, B. Duncan, X. Lascelles, Allan Lepine, Sally Perea, et al., "Aging in Cats: Common Physical and Functional Changes," *Journal of Feline Medicine and Surgery* 18, no. 7 (2016): 533–550; Dan G. O'Neil, David B. Church, Paul D. McGreevy, Peter C. Thomson, and David C. Brodbelt, "Longevity and Mortality of Cats Attending Primary Care Veterinary Practices in England," *Journal of Feline Medicine and Surgery* 17, no. 2 (2015): 125–133; Alice Villalobos, "Cancers in Dogs and Cats," in *Hospice and Palliative Care for Companion Animals*, ed. Amir Shanan, Jessica Pierce, and Tamara Shearer (Hoboken, N.J.: Wiley, 2017), 89–100; Renaud Kaeuffer, Dominique Pontier, Sébastien Devillard, and Nicolas Perrin, "Effective Size of Two Feral Cat Populations (Felis Catus L.): Effect of the Mating System," *Molecular Ecology* 13, no. 2 (2004): 483–490.

19. Myron and Witter, *Dewey*, 203–208; Lynn A. Planchon, Donald I. Templer, Shelley Stokes, and Jacqueline Keller, "Death of a Companion Cat or Dog and Human Bereavement: Psychosocial Variables," *Society & Animals* 10, no. 1 (2002): 93–105; Claudia Pilatus and Gisela Reinecke, *Es ist doch nue ein Hund . . . Trauer um Tiere* (Nerdlen: Kynos, 2008); Frantz Cappé, *Mon Chat, mon chien va partir* (Paris: Albin Michel, 2017).

20. Maria B. Rosenkränzer, *Lilly, unser kleine Tiger* (Frankfort: Fischer, 2006).

21. James Bowen, *A Street Cat Named Bob* (London: Hodder & Staughton, 2012).

22. Frances Simpson, *The Book of the Cat* (London: Cassel, 1903), 32, 34; Éric Pierre, "Amour des hommes—Amour des bêtes" (doctoral thesis, Angers, 1998).

23. Paul Léautaud, *Journal littéraire* (Paris: Mercure de France, 1986), 1:1077, 1399, 1822, 2:1118, 1359–1360, 1773, 3:165, 171, 1357; Léautaud, *Journal littéraire: Histoire du journal*, 109, 134, 150, 155–156.

24. Janet M. and Steven F. Alger, *Cat Culture: Philadelphia: The Social World of a Cat Shelter* (Philadelphia: Temple University Press, 2002).

Chapter 8. Catdogs

Jonah

1. Helen Brown, *Cleo: The Cat Who Mended a Family* (Crows Nest: Allen & Unwin, 2009)

2. Helen Brown, *After Cleo Came Jonah: How a Crazy Kitten and a Rebelling Daughter Turned Out to Be Blessings in Disguise* (Crows Nest: Allen & Unwin, 2012). (The reader should note that we are referencing the American edition of Helen Brown's book, which bears a different title, *Cats and Daughters: They Don't Always Come When Called* [New York, Citadel Press, 2012]—T. N.].)

3. Brown, *Cats and Daughters*.

4. Brown, *Cats and Daughters*, 7, 104, 106, 149.

5. E. Larieux and Ph. Jumaud, *Le Chat: Races—elevage—maladies* (Paris: Vigot Frères, 1926), 33; Dennis C. Turner and Patrick Bateson, eds., *The Domestic Cat* (Cambridge: Cambridge University Press, 2014), 17–18, 23–24, 28–31, 140; Rachel A. Casey, Sylvia Vandenbussche, John W. S. Bradshaw, and Margaret A. Roberts, "Reasons for Relinquishment and Returns of Domestic Cats (Felis sylvestris catus) to Rescue Shelters in the U.K.," *Anthrozoös* 22, no. 4 (2009): 347–358; Mathieu Harel, Mathilde Goujon, and Dominique Grandjean, "Contribution à l'étude des refuges félins en France" (veterinarian thesis, Maisons-Alfort, 2014), 121.

6. Brown, *Cats and Daughters*, 111–116; Brown, *Cleo*, 16.

7. Brown, *Cats and Daughters*, 113, 116–117, 123, 143.

8. Jean-M.-M. Rédarès, *Traité raisonné sur l'éducation du chat domestique* (Paris: Bourayne, 1835), 77; Gaston Percheron, *Le Chat* (Paris: Didot, 1885), 129; Larieux and Jumaud, *Le Chat*, 40–49; Turner and Bateson, *Domestic Cat*, 17, 23–24.

9. Harel, Goujon, and Grandjean, "Contribution à l'étude des refuges"; Kathryn Dybdall and Rosemary Strasser, "Is There a Bias against Stray Cats in Shelters?" *Anthrozöos* 27, no. 4 (2014): 603–614.

10. Brown, *Cats and Daughters*, 109; Turner and Bateson, *Domestic Cat*, 144, 160; Marta Borgi and Francesca Cirulli, "Children's Preferences for Infantile Features in Dogs and Cats," *Human-Animal Interaction Bulletin* 1, no. 2 (2013): 1–15; *Journal of Feline Medicine and Surgery* 17, no. 9 (2015); V. Santoire et al., ed., "Stérilisation félin," special issue, *PratiqueVet*, 2017, suppl. 152.

11. Scott R. Loss and Peter P. Marra, "Population Impacts of Free-Ranging Domestic Cats on Mainland Vertebrates," *Frontiers in Ecology and the Environment* 15, no. 9 (2017): 502–509.

12. Linda C. Marston and Pauleen C. Bennett, "Admissions of Cats to Animal Welfare Shelters in Melbourne, Australia," *Journal of Applied Animal Welfare Science* 12, no. 3 (2009), online; Samia R. Toukhsati, Emily Young, Pauleen C. Bennett, and Grahame J. Coleman, "Wandering Cats: Attitudes and Behaviors towards Cat Containment in Australia," *Anthrozoös* 25, no. 1 (2012): 61–74.

13. Turner and Bateson, *Domestic Cat*, 14.

14. Ibid., 70, 152–153, 217–227; Alice Bouillez, "Problèmes des chats errants et gestation de ces populations" (veterinarian thesis, Lyon, 2015); "Chats errants en France," dossier *One Voice*, 2018.

15. Jennifer L. McDonald, Mairead Maclean, Matthew R. Evans, and Dave J. Hodgson, "Reconciling Actual and Perceived Rates of Predation by Domestic Cats," *Ecology and Evolution* 5, no. 14 (2015): 2745–2753.

16. Turner and Bateson, *Domestic Cat*, 188.

17. Brown, *Cats and Daughters*, 1; Brown, *Cleo*, 46, Turner and Bateson, *Domestic Cat*, 144–145; Marcus Schneck and Jill Caravan, *Le Chat* (Paris: Solar, 1991), 51, 58; Honoré de Balzac, *The Human Comedy: Selected Stories*, trans. Linda Asher (New York: New York Review of Books Classics, 2014).

18. Marcel Mauss, "Les Techniques du corps," *Journal de la Psychologie* 32, nos. 3–4 (1936): 271–293; Erving Goffman, *The Presentation of Self in Everyday Life* (New York: Anchor, 1959); Alain Corbin, *The Foul and the Fragrant: Odour and the Social Imagination* (New York: MacMillan, 1994).

19. Denys Cuche, *La Notion de culture dans les sciences sociales* (Paris: La Decouverte, 2004); Tim Ingold, *Marcher avec les dragons*, trans. Pierre Madelin (Brussels: Zones sensibles, 2013); Janet M. and Steven F. Alger, *Cat Culture: Philadelphia: The Social World of a Cat Shelter* (Philadelphia: Temple University Press, 2002); Dominique Guillo, "Les Récherches éthologiques récentes sur les phéromones socio-culturels dans le monde animal," *L'Année sociologiques* 66, no. 2 (2016): 351–384; Florent Kohler, "Comment les émotions font corps," in *Croiser les sciences pour lire des animaux*, ed. Éric Baratay (Paris: Editions de la Sorbonne, 2020) 269–286; Elizabeth Marshall Thomas, *The Tribe of Tiger: Cats and Their Culture* (London: Weidenfield and Nicolson, 1994).

20. Brown, *Cats and Daughters*, 113–116, 201;

21. Ibid., 120, 136, 182, 199; Turner and Bateson, *Domestic Cat*, 18, 238; Dominique Lestel, *Les Amis de mes amis* (Paris: Seuil, 2007); David Goode, *Playing with My Dog Katie* (West Lafayette, Ind.: Perdue University Press, 2006).

22. Brown, *Cats and Daughters*, 157.

23. Ibid., 142–143.

24. Ibid., 145, 157; Dennis Burnham, Elizabeth Francis, U. Vollmer-Conna, C. Kitamura, Vicky Averkiou, A. Olley, M. Nguyen, and Cal Paterson, "Are You My Little Pussy-Cat? Acoustic, Phonetic, and Affective Qualities of Infant- and Pet-Directed Speech," 5th International Conference on Spoken Language Processing, Sydney, 1998.

25. Brown, *Cats and Daughters*, 149.

26. Ibid., 116–117.

27. Turner and Bateson, *Domestic Cat*, 188–199; Colleen Wilson, Melissa Bain, and Gary Landsberg, "Owner Observations Regarding Cat Scratching Behavior," *Journal of Feline Medicine and Surgery* 18, no. 10 (2016): 791–797; Gary M. Landsberg, "Cat Owners' Attitudes toward Declawing," *Anthrozoös* 4, no. 3 (1991): 192–197; Marta Amat, Tomàs Camps, and Xavier Manteca, "Stress in Owned Cats: Behavioural Changes and Welfare Implications," *Journal of Feline Medicine and Surgery* 18, no. 8 (2016): 577–586.

28. Brown, *Cats and Daughters*, 260–261.

29. Ibid., 236, 244, 262, 321.

30. Amat, Camps, and Manteca, "Stress in Owned Cats"; Catherine Mège, Claude Béata, Gérard Muller, Colette Arpaillange-Vivier, Edith Graff, Dominique Lachapèle, Muriel Marion, Nathalie Marlois, and Nicolas Massal, *Pathologie comportementale du chat* (Paris: Masson, 2012).

31. Albert Camus, *The First Man*, trans. David Hapgood (New York: Vintage, 1992), 130–140.

End of the Journey

1. Helen Brown, *Cats and Daughters: They Don't Always Come When Called* (New York: Citadel Press, 2012), 316–317.

INDEX

ANIMAL VOICES, ANIMAL WORLDS

Erin McKenna, *Livestock: Food, Fiber, and Friends*

Arnold Arluke and Andrew Rowan, *Underdogs: Pets, People, and Poverty*

Michele Merritt, *Minding Dogs: Humans, Canine Companions, and a New Philosophy of Cognitive Science*

Anne Benvenuti, *Kindred Spirits: One Animal Family*

Éric Baratay, *Animal Biographies: Toward a History of Individuals*

Éric Baratay, *Feline Cultures: Cats Create Their History*